Essentials of Sea Survival

Frank Golden, MD, PhD

Michael Tipton, PhD

Human Kinetics

Library of Congress Cataloging-in-Publication Data

Golden, Frank, 1936-
   Essentials of sea survival / Frank Golden, Michael Tipton.
     p. cm.
  Includes bibliographical references and index.
   ISBN 0-7360-0215-4
   1. Survival after airplane accidents, shipwrecks, etc.--Handbooks,
manuals, etc. I. Tipton, Michael, 1959- II. Title.
   VK1259 .G65   2002
   613.6'9--dc21

                           2002022911

ISBN-10: 0-7360-0215-4

ISBN-13: 978-0-7360-0215-8

Acquisitions Editor: Michael S. Bahrke, PhD; **Developmental Editors:** Jim Murphy, Rebecca Crist; **Assistant Editor:** Mark E. Zulauf; **Copyeditor:** Bob Replinger; **Proofreader:** Julie A. Marx; **Indexer:** Susan Hernandez; **Permission Manager:** Dalene Reeder; **Graphic Designer:** Fred Starbird; **Graphic Artist:** Sandra Meier; **Photo Manager:** Les Woodrum; **Cover Designer:** Keith Blomberg; **Photographer (cover):** nordicphotos.com/Vigfus Birgisson; **Art Managers:** Kelly Hendren and Carl Johnson; **Illustrators:** Roberto Sabas and Accurate Art; **Printer:** United Graphics

Human Kinetics books are available at special discounts for bulk purchase. Special editions or book excerpts can also be created to specification. For details, contact the Special Sales Manager at Human Kinetics.

Printed in the United States of America

Printed in the United States of America   10 9 8 7 6 5 4 3

Human Kinetics
Web site: www.HumanKinetics.com

United States: Human Kinetics
P.O. Box 5076
Champaign, IL 61825-5076
800-747-4457
e-mail: humank@hkusa.com

Canada: Human Kinetics
475 Devonshire Road, Unit 100
Windsor, ON N8Y 2L5
800-465-7301 (in Canada only)
e-mail: info@hkcanada.com

Europe: Human Kinetics
107 Bradford Road
Stanningley
Leeds LS28 6AT, United Kingdom
+44 (0)113 255 5665
e-mail: hk@hkeurope.com

Australia: Human Kinetics
57A Price Avenue
Lower Mitcham, South Australia 5062
08 8372 0999
e-mail: info@hkaustralia.com

New Zealand: Human Kinetics
Division of Sports Distributors NZ Ltd.
P.O. Box 300 226 Albany
North Shore City, Auckland
0064 9 448 1207
e-mail: info@humankinetics.co.nz

This book is dedicated to all those who lost their lives at sea and the many survivors whose accounts, sometimes given at the cost of resurrecting unpleasant memories, have provided us with the important anecdotal evidence to both stimulate research and support its conclusions.

Contents

Preface

" **A**T LAST SOME SURVIVORS were sighted drifting in an open boat. Their ship had been sunk seventy hours previously, but it had taken three days for the rescue ship to locate them in the cold bleak Northern waters. In the poor visibility, it could just be discerned that at least some were alive. What of the remainder? On coming alongside it quickly became obvious that only a fraction of the crew had managed to board the lifeboat, and of those only a handful had survived the time adrift. Even then, they were in an appalling condition; the cold had taken its toll—some were semiconscious, while all would be lucky to leave the hospital with all four limbs intact."

This extract is from the report of the sinking of a merchant ship on an Arctic convoy in the early days of World War II. It was one of the first incidents to suggest that, among other medical conditions, the clinical effects of exposure to cold would be a major problem confronting survivors in the succeeding years of the conflict. Little was known about the underlying mechanisms causing these problems or the best ways to treat them. After all, as the opening chapter of this book will show, our ancestors had accepted death at sea as an occupational hazard for thousands of years, ever since they first took to the seas to fish, trade, explore, and make war. Even the introduction of the cork life jacket by John Ross Ward in 1851, the first real innovation in survival devices, could do little to stem the tide of maritime casualties.

In World War II the alarming loss of life at sea in the survival phase (following ship abandonment) prompted the British naval authorities to ask the Medical Research Council to appoint a special "Shipwreck Committee" to inquire into the physiological and medical issues involved. The committee collated the accounts of survivors and rescuers, identified areas needing research, and produced practical guidelines on both treatment of such casualties and preventive measures to improve survival prospects (War Memorandum No. 8: *A Guide to the Preservation of Life at Sea after Shipwreck* 1943).

At the end of the war, the Admiralty set up a committee under Rear Admiral A. G. Talbot RN to investigate the causes of the loss of life of naval personnel at sea in the course of the war, together with all aspects of naval lifesaving. The findings of the committee included the startling statistic that over two-thirds (approximately 30,000) of all Royal Navy fatalities during the war occurred during the survival phase. The conclusions of the committee stimulated the research and development that resulted in modern life jackets (personal flotation devices), inflatable life rafts, and sea survival rations.

In the intervening years, exciting research in open-water survival has continued internationally, albeit spasmodically. This work, together with technological advancements, has done much to improve our overall understanding of the problems associated with immersion in cold water and the ability to survive at sea. Yet, worldwide, many thousands of lives continue to be lost in this manner every year. Why do these catastrophes continue to occur?

Analyses of maritime tragedies suggest two principal underlying causes. First, there appears to be a general lack of understanding of the nature of the various threats and the reaction of the body *(physiological responses)* to them. Consequently, policies, regulations, specifications, and design of protective equipment often fall short of the ideal. Second, in a survival situation, costly safety equipment is often not readily at hand, is difficult to operate in adverse conditions, or is impossible to use correctly without specific training. Survivors are often left to their own devices to adapt to the situation as best they can.

If you frequently participate in aquatic activities, have you stopped to consider the nature of the threat confronting you? What steps can you take to improve your chances of surviving in the water or a survival craft? Is your knowledge of the basic principles involved sufficient to enable you to adapt, improvise, and thereby survive?

Drawing upon historical anecdotes as well as published scientific research, this book examines the nature of the many threats confronting the survivor at sea and outlines, in lay terms, the methods to prevent or minimize the dangers. The first half of the book deals predominantly with the physiological and behavioral responses to cold, immersion, and drowning. The second half addresses techniques for survival and rescue, either in the water or in a lifeboat. The lessons it teaches are drawn from classical maritime disasters and personal accounts of near-miraculous survival, as well as carefully controlled laboratory experiments. In order to make the text more readable, it is not heavily referenced, although a bibliography is provided. The aim of the authors is to provide both the specialist and the layperson with a compelling, informative, and comprehensive guide to open-water survival, particularly in cold conditions. Having read the book, you will have a comprehensive and practical understanding of the dangers you face. Consequently, you then should be able to construct an informed survival strategy and be able to maximize your chance of open-water survival should disaster strike. Thus, we hope you will never be among the 140,000 people who die each year in water-related incidents worldwide.

Acknowledgments

No book on this topic would be complete without acknowledging the many contributions made by so many dedicated scientists and their associates through the years. We are proud to have so many of them as friends. For these authors, two of our predecessors are worthy of a special mention and a vote of thanks: Professors McCance and Hervey. Both were Chairmen of the Royal Navy Personnel Research Committee and, in their time, made outstanding personal contributions in this area, as well as providing the stimulation and encouragement to others, including the authors in our earlier days. We, and many seafarers, owe them much.

We would especially like to record our unstinting admiration for the bravery, stoicism, and dependability of the many volunteers who acted as experimental subjects for us in the past. We hope we have done them justice. We would also like to acknowledge the support, encouragement and helpful suggestions provided by those we have worked with over the years, in particular Chief Petty Officer Pete Moncaster, Royal Navy, for his technical support in our "formative years." He did everything except succeed in walking on water!

Our thanks are also due to the editorial staff of Human Kinetics and to colleagues who provided helpful advice on the initial draft of this book.

Finally, a heartfelt "thanks" to our wives and families whose sacrifices, patience, and understanding have enabled this book to sail the sometimes rough, sometimes chilly, waters to its final destination.

Credits

Figure 1.1 – Reprinted with kind permission from the MV Estonia. Final Report, 1997.

Figure 1.2 – Reprinted with kind permission of "The Age" Newspaper, Melbourne, Australia.

Figure 4.02 – Reprinted with permission from Elsevier Science (*The Lancet*, 1955, Vol.2, 761-768).

Figure 4.04 – Adapted, by permission, from St. F.C. Golden and M.J. Tipton. 1988, "Human adaptation to repeated cold immersions," *Journal of Physiology* 396:349–363.

Figure 6.01 – Adapted, by permission, from F. Golden, 1973, "Recognition and treatment of immersion hypothermia," *Proceedings of the Royal Society of Medicine* 66:1058-1061.

Figure 6.05 – Adapted, by permission, from F. Golden, 1973, "Recognition and treatment of immersion hypothermia," *Proceedings of the Royal Society of Medicine* 66:1058-1061.

Figure 7.01 – Reprinted, by permission, from G.W. Molnar, "Survival of hypothermia by men immersed in the ocean," *Journal of the American Medical Association* 131:1046-50.

Figure 7.03 – © Crown Copyright 1987/ MOD. Reproduced with the permission of the Controller of Her Majesty's Stationary Office.

Figure 7.06 – Reprinted, by permission, from J.S. Hayward, J.D. Eckerson, and M.L. Collis. 1975, "Effect of behavioral variables on cooling rate of man in cold water," *Journal of Applied Physiology* 38:1073–1077.

Figure 11.3 – Reprinted, by permission, from St. F.C. Golden and G.R. Hervey, 1977, "The mechanism of the afterdrop following immersion hypothermia in pigs," *Journal of Physiology* 272:26–27.

1

A Catalog of Disaster

"SOME WERE DASHED TO PIECES** on the rocks, others drowned, while others died from cold . . ."

Herodotus, Histories Book VI, circa 450 B.C.

The preceding quotation is a reminder that nothing is new in the history of survival at sea. A fatality occurs almost every day somewhere in the world. Such incidents are so common that most receive no more than a small mention in a local newspaper. Only when a large loss of life occurs or when a celebrity is involved does the news hit the front page. In most cases, the survival lessons are not considered or are soon forgotten, condemning later generations to suffer the same experience. In this chapter we review some of the better-known examples of the recent past so that we can identify and prioritize the areas most pertinent to survival in open water.

Few who go to sea, either professionally on a frequent schedule or for leisure on an occasional basis, ever think that they might become survivors.[1] Most believe it will never happen to them. Nevertheless, the prudent will carefully consider the matter and periodically review their preparations. In the United Kingdom, one in every four people who use the sea for recreational purposes has been involved in a life-threatening situation (Royal National Lifeboat Institution statistics 1999). In the United States in the 10 years between 1985 and 1995, there were 204 boat collisions, 769 groundings, 438 "strikings," 211 fires, 131 "sinkings," 38 capsizes, and 18 explosions (*USA Today* 11 January 1999).

---

1. Throughout this book we use the term *survivor* to describe all who find themselves in a survival situation regardless of whether they ultimately survived.

In reply to the question, "What do you think are the major threats or problems involved in surviving at sea?" many will reply, "Drowning." This is technically correct but only part of the story. Drowning may be the end of a survival saga that frequently starts with the coincidental occurrence of a series of unrelated incidents that culminate in an "accident." Knowledge of the process and strategies for preventing its progression can be critical to the outcome.

Many consider the major problems confronting a survivor at sea as a lack of water or food. These views reflect media coverage of epic survival voyages that dominate the headlines from time to time. Such stories usually relate to survivors who were adrift, often for weeks, in an open boat or life raft in tropical waters. The accounts of the privations they endured make compelling reading, much more fascinating than the account of someone who spent a day adrift in temperate waters. Yet it is entirely possible that the latter may have endured the greater hardship and come closer to death. Such perilous experiences occur regularly among recreational sailors and fishers yet never make the headlines because only one or two people are involved or because they do not

The prudent person will consider the possibility of an accident, along with strategies for survival.

Official U.S. Coast Guard photograph.

survive to tell the tale. Thus, misconceptions persist regarding the nature of the real threat.

SHIP SURVIVAL SAGAS

Between 1978 and 1998, more than 5,300 passengers were killed in ferry accidents around the world, making ferry travel 10 times more dangerous than air travel (Faith 1998). Most people are familiar with the story of the SS *Titanic,* popularized in an Oscar-winning Hollywood film, but the record to date for the number of lives lost at sea in a single peacetime incident goes to the sinking of the passenger ferry *Dona Paz,* after a collision with a small oil tanker in the Philippine archipelago, on a calm moonless night on 28 December 1987. Of the estimated 3,156 people on board the ferry, none of the crew and only 24 passengers survived. The remainder were apparently unable to evacuate the sinking burning ship because of blazing oil on the surrounding water. It was reported that sharks attacked some who had managed to survive the fire. Two of the tanker's crew of 13 survived.

Larger losses of life occurred in single incidents in wartime, but exact numbers lost are difficult to establish because of censorship and the absence of precise passenger manifests. In June 1940 German aircraft off St. Nazaire in western France sank the British liner RMS *Lancastria* with an estimated loss of more than 6,000 lives. Most of those killed were troops evacuating France, although a large number of civilians (women and children) also died. The captain survived but was later to lose his life when another ship under his command, *Laconia,* was torpedoed in September 1942 with the loss of 5,200 lives. In January 1945 the German liner *Wilhelm Gustloff* was torpedoed and sunk in the Baltic Sea off Danzig by a Russian submarine, with an estimated loss of over 10,000 lives. In the Pacific many Japanese troop ships, including some ferrying prisoners of war, were sunk with huge losses of life.

In the same year that *Dona Paz* was lost, another ferry, *Herald of Free Enterprise,* was wrecked just outside Zeebrugge, Belgium. In this incident the ship capsized and was lying, partially flooded, on its side on a sand bank, surrounded by rescue ships. Even in those comparatively favorable circumstances, many people failed to escape because of entrapment or through incapacitation by cold or injury. At least 193 of the 580 on board perished. Because of the absence of compartmentalization on their cargo decks, car ferries have proved to be inherently unsafe.

A recent tragedy involving another ferry was the loss of the MV *Estonia.* She sank in the Baltic Sea south of Finland on 28 September 1994. Several points from this incident highlight the problems that people are likely to encounter in similar survival situations. Perhaps the most important of these is the speed with which disaster can unfold, leaving little time for innovative thinking or reading safety instructions.

As with the Canadian Pacific liner *Empress of Ireland,* which sank in 14 minutes in the icy waters of the St. Lawrence river in May 1914 with the loss of over 1,057 lives, only a short time was available for evacuation. From the moment *Estonia* first began to take on a serious list and it

Estonia

The report of the official inquiry into the *Estonia* disaster (1997; see figure 1.1) records that the ship "departed from Tallinn, the capital of Estonia, on the 27 Sep 1994 at 1915 for a scheduled voyage to Stockholm. She carried 803 passengers and 186 crew (total 989). As the voyage continued the wind velocity increased gradually to about 18–20 m/s and veered to the South West, producing a significant wave height[2] of 3–4m on the port bow. At 0100 a seaman, doing a scheduled round of the car deck, heard a loud metallic bang from the bow area as the vessel hit a large wave. He reported what he heard but on inspection could find no damage. Periodically thereafter the banging was heard by a variety of people but no cause could be detected. About 15 minutes after (0115), the bow visor separated from the bow and tilted over the stem. The ramp was forced fully open allowing large volumes of water to enter the car deck. Very rapidly the ship took on a heavy starboard list, turned to port and slowed down. Passengers started to rush up the staircases and panic developed in many places. Many passengers were trapped in their cabins and had no chance of getting out in time. Lifejackets were distributed to those passengers who managed to reach the boat deck. They jumped, or were washed, into the sea. Some managed to climb into liferafts, which had been released from the vessel. No lifeboats could be launched because of the heavy list.

The first Mayday was transmitted at 0122 and last contact was at 0129. The ship by now was lying on its starboard side; it was submerged by 0130. It sank rapidly stern first and disappeared from the radar screens of nearby ships at 0150.The first rescue helicopter arrived on scene at 0305.

During the night and early morning, helicopters and ships rescued 138 people, of whom one later died in hospital. A further 92 dead bodies were recovered."

2. Significant wave height (HSig) is the term given to the average height of the top 33 percent of waves. About one in every hundred waves will reach twice the significant wave height; for example, if HSig = 4 meters (13 feet) then HMax = 8 meters (26 feet).

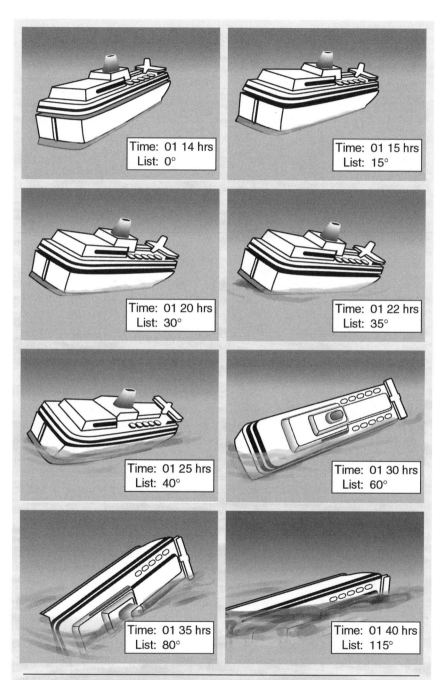

Time: 01 14 hrs
List: 0°

Time: 01 15 hrs
List: 15°

Time: 01 20 hrs
List: 30°

Time: 01 22 hrs
List: 35°

Time: 01 25 hrs
List: 40°

Time: 01 30 hrs
List: 60°

Time: 01 35 hrs
List: 80°

Time: 01 40 hrs
List: 115°

Figure 1.1 Computer-generated pictures of *Estonia* sinking.
Reprinted from the MV *Estonia,* Final Report 1997.

became apparent that she was in trouble, the time for evacuation would appear to have been 10 to 20 minutes. The description of those final minutes is one of an all-prevailing sense of confusion and panic. In the violently moving ship, lying at an acute angle, people struggled to make their way onto the open decks. Stairs and bulkheads were at unhelpful attitudes. Doors, designed to work when vertical, were difficult to open, while unrestrained furniture, gaming machines, and so forth fell about, injuring or trapping people. About 300 managed to make their way to the open deck and then struggled to maintain their balance and hold their position long enough to put on life jackets that were being distributed from containers. This task was difficult because people were unfamiliar with the correct method of donning and securing the life jackets. Consequently, on entering the water, many of the life jackets either came off or became displaced, rendering them useless. Less than half of those who made it to the upper deck survived.

Because of the severe list, lifeboats could not be launched. Many people tried to launch inflatable life rafts, but again their unfamiliarity with the equipment and the list of the ship made this extremely difficult. The rafts that were successfully launched ended up in positions that made subsequent boarding difficult. In the end, with the rapid sinking of the ship, the majority of survivors found themselves in the dark, being tossed about by the waves, surrounded by a mass of unused life jackets, partially inflated life rafts, and damaged or upturned lifeboats. Many people were shouting for help. Some swam and boarded life rafts only to find that they were inverted and therefore offered little protection from the environment, apart from providing a partially safe refuge from drowning.

Those who managed to board rafts that were the right way up often found that the canopy support arches had not fully inflated, probably because water from the breaking waves had weighed them down before complete inflation had occurred. To make matters worse, waves washed in and out of the rafts, so they offered little or no protection from the cold water. Also, the occupants of the rafts had to hold on to something to avoid being buffeted about or washed back into the sea. Only with difficulty could they help other survivors outside the rafts, who were holding on to ropes or lanyards, or trying to clamber in. Many were impeded in boarding because their legs or arms had become entangled in the myriad of loose lines near the rafts. For the majority, the physical effort was too much, and after several unsuccessful attempts they became exhausted and gave up.

The occupants of some rafts were able to help a few exhausted, cold people to board, only to see them immediately washed out by the next wave. After a while the cries for help from those in the water diminished, although, remarkably, one German man swam for three hours and successfully managed to attract the attention of rescuers with a flashlight!

Survival was not guaranteed even for those who managed to remain in the rafts or on upturned lifeboats. Many worked hard to try to bail the water out of the rafts. Regrettably, because of unfamiliarity and cold hands, they could not close the canopy apertures correctly; consequently, the waves washed more water into the raft than the occupants were capable of bailing. In many rafts the survival-aid container had been washed out before the occupants managed to board the raft. In rafts where the containers were still present, some survivors described how they were unable to open them to get at the bailer because of cold hands. The plastic wrappings of the flares proved equally challenging to cold fingers. One survivor described how, driven by frustration, he tried using his teeth but gave up when he pulled some out! Other survivors were successful in using the flares, but some, not sure how to use them, inadvertently discharged them inside the raft. Some tried to don emergency thermal protective devices found in the rafts, but these proved too flimsy and tore. Eventually, with the passage of time and with most inadequately clothed for the conditions, many lost consciousness through hypothermia and were washed overboard. Others simply slid into the water that was swilling around in the bottom of the partially flooded rafts, and drowned.

Other passenger ferries in the vicinity arrived on the scene fairly quickly—the first in about 45 minutes—but because of their high freeboard and the prevailing sea state, they were unable to launch boats. Consequently, they could do little about rescuing survivors from the water. One did manage to lower and recover a life raft manned with some crew members to assist survivors. Unfortunately, when another ship attempted a similar maneuver, on recovery the floor of the raft ripped just as it reached deck level, spilling the occupants back into the sea, where many drowned.

Most of those who survived were rescued by some of the many helicopters that arrived on scene within a few hours of the sinking, the first about one and one-half hours after the ship capsized. Regrettably, even this form of rescue was not without problems. Initially it was difficult to locate survivors in the dark. The fact that none of the life jackets had lights contributed to the difficulty. Some helicopters suffered winch malfunctions, and one had an engine problem that forced it to return to base. Most rescue men operating at the end of the winch wire became fatigued; they had the arduous job of helping extremely cold, incapacitated survivors while being buffeted by waves. Some of the rescue men were injured in collisions with the rolling inverted lifeboats while searching for casualties. All survival craft had to be searched. Rescuers wasted valuable time examining craft that crews from other helicopters had already searched, an unnecessary activity that added to crew fatigue. Most of the successful rescues occurred within four or five hours of the sinking. After six to seven hours no more survivors were found.

The report does not mention air or water temperature, yet at autopsy, in all cases of drowning, hypothermia was given as a contributory factor (the sea temperature was approximately 12 degrees Celsius; 54 degrees Fahrenheit). Of the bodies autopsied, it was noted that 25 were naked, 18 had very insufficient clothing, 40 had insufficient clothing, and only 10 wore extra clothing.

Perhaps not surprisingly, given the physical difficulties in abandoning ship and the severity of the subsequent survival ordeal, the majority of those rescued were males, aged 15 to 44. Only 3 percent of males over 65 survived. No females over 65 survived. Only 5 percent of the females on board survived compared with 22 percent of the males (table 1.1).

Table 1.1   Breakdown of Passengers and Crew of the MV *ESTONIA* by Gender and Number (Percentage) of Survivors

	Passengers			Crew			Total
	Male	Female	Total	Male	Female	Total	
Total	418	385	803	86	100	186	989
Survivors (%)	80(19)	14(4)	94(12)	31(36)	12(12)	43(23)	137(14)

YACHTING SURVIVAL SAGAS

On a smaller scale, some of the more dramatic survival stories in recent years have emanated from the leisure-sailing world. One of the most notable of these was the 1979 Fastnet race. This annual 600-mile race off the south coast of the British Isles ended in tragedy in August of that year. An unexpected Atlantic front produced a severe storm that confronted the 303 entrants with winds of about 60 to 70 knots (force 11–12). The confused sea state intensified the problem, resulting in the abandonment of 19 yachts (although 14 were later recovered!) and the sinking of 5. The rescuers recovered 136 men and women. Sadly, 15 lives were lost. Seven died after safely boarding their life rafts, and another 3 died during rescue.

The 1998 Sydney-to-Hobart ocean sailing race ended in similar circumstances (Mundle 1999). This event is claimed to be one of the most prestigious on the ocean sailing calendar. The crossing of the unpredictably treacherous Bass Straight is considered a challenge that will "sort out the men from the boys!" Perhaps because of that fearsome reputation, those who enter are highly competent ocean sailors fully aware of the possible threat and the risks they are undertaking; accordingly they are properly equipped. The race statistics tend to support this: In its 54-year history before 1998, just two lives had been lost. These statistics were about to change as the 115 entrants left Sydney harbor on 26 December 1998.

The day started sunny and calm, although the weather forecast warned of a deep depression with southerly winds up to 50 knots for late afternoon and evening, when most of the entrants would be near the Bass Straight. Twenty hours into the race the storm struck with sudden ferocity. Gusts of 80 miles per hour were recorded. Some crews described the seas as being "as high as three-story buildings." Radio altimeter readings and observations from helicopters confirmed those estimates, reporting waves as high as 18 meters (60 feet) with a further 6 meters (20 feet) of curling, breaking crest on top of some waves. The Coast Guard recorded that over 80 distress flares had been sighted in the space of minutes. Crews broadcast dozens of Mayday calls. Full search and rescue (SAR) actions were immediately implemented. At the height of the rescue as many as 25 aircraft were on scene.

The huge waves crashing over the boats washed some crew out of their cockpits and overboard. Other boats suffered dismasting and structural damage. Some yachts were simply knocked down repeatedly. Others rolled over but fortunately managed to self-right. Sadly, a British Olympic yachtsman was lost when his safety-harness lanyard broke as his yacht rolled through 360 degrees after a huge wave hit it. Some who had been washed out of the cockpit, or overboard, were more fortunate. Their harnesses remained intact, enabling them to struggle back on board. An American survivor described how "the sea just crashed into the cockpit and I got washed out before I got a chance to attach my lifeline!" Fortunately, he was located 40 minutes later by a police helicopter using infrared equipment, and a dramatic rescue ensued.

The enormous seas, with the highly disturbed surface conditions, made it extremely difficult to locate anyone in the water or conduct a rescue, even from an undamaged boat. Crews in boats relatively close to each other frequently failed to make visual contact. Most of the bruised, exhausted competitors ran for shelter in their damaged boats, thankful to have survived the ordeal.

Seventy of the 115 yachts starting the race retired. Fifty-five crew members were evacuated from 6 yachts that were abandoned—47 by helicopter, 6 by fishing boat, 2 by a Royal Australian Navy warship. Six of the original 9 occupants of one life raft were rescued. Given the horrendous circumstances, it is a tribute to the professionalism of all involved that only 6 lives were lost. The SAR authorities in particular deserve a special commendation in this respect.

Other recent dramatic attempts to rescue lone sailors in difficulties in huge seas in the Southern Ocean have also provided exciting television viewing for those sitting in the comfort of their armchairs. In this respect, the events of the 1996–97 Vendee Globe were particularly enthralling. The account by Peter Goss (1998) of his rescue of a fellow competitor makes compelling reading (see also chapter 11). Having endured several knockdowns in atrocious conditions and fighting for survival himself, he selflessly, and without question, turned around into hurricane-force winds to sail 257 kilometers (160 miles) back to the assistance of another competitor, Raphael Dinelli. The Frenchman was drifting in a small life raft in huge seas, hundreds of miles from possible rescue by other sources. With the help of a Royal Australian SAR aircraft, Goss eventually located the lone survivor and managed to rescue him. About two weeks later in the same race, two other competitors needed rescuing—Thierry Dubois, who was marooned on the hull of his upturned yacht, and Tony Bullimore, who spent four days inside his inverted yacht (figure 1.2; see also chapter 9).

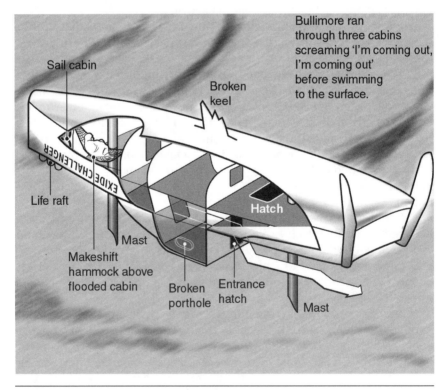

Figure 1.2   Diagram of Tony Bullimore's upturned yacht.

Reprinted from *The Age* newspaper, Melbourne, Australia.

Besides highlighting the heroism and determination of the individuals involved, these survival accounts also demonstrate the need for survival strategies, survival equipment, clothing, training, and rescue skills. Such strategies are based on accurate identification of the likely nature of the physiological threat confronting the survivor and the criteria to be met in combating that threat. This knowledge has been accumulated over the years, from actual incidents and unobtrusive experiments. Since the end of World War II, these experiments have been conducted with the help of volunteers in laboratories in many parts of the world. Without such experiments, progress would have been absent, or at best pitifully slow.

Sadly, for every successful survival story highlighted each year by the media, many hundreds of people die needlessly at sea in spite of the technological progress that has been made. It may be surprising to many to discover that the most hazardous occupation in the United States today is commercial fishing (Finkel 1999). Commercial inducements and a limited season frequently create intolerable working pressure to secure a share of the potentially rich harvest. In recent years in Alaska, the small community of snow-crab fishers has suffered almost one fatality a week during the relatively short season, and the personal injury rate runs at 100 percent! British and Canadian fishing statistics are similarly depressing. Decreasing fish stocks coupled with market pressures are forcing many fishers to take risks and remain fishing in adverse conditions that previously they would have avoided.

ANATOMY OF AN ACCIDENT

Many accidents,[3] both large- and small-scale, result from a similar set of circumstances. Usually a number of factors are involved, any one of which would have little effect but which in combination result in a deadly spiral of events leading to disaster. This coincidental combination of events may be a mixture of human omissions, errors of judgment, inadvertent actions, poor communication, and adverse weather together with a host of contributory personal and extraneous factors. Therefore, vital components of a good survival strategy are recognition of the warning signs and awareness of the most appropriate corrective response to avert disaster. Accomplishing this requires knowledge, experience, and avoidance of denial.

---

3. The term *accident* is inappropriate to the majority of disasters at sea because rarely do they occur by chance or without apparent cause. Instead, they are the result of negligence or disregard of the premonitory warning signs through casualness, ignorance, overconfidence, or poor operating procedures.

Titanic

One of the best-documented maritime examples of such an "accident" is that of the "unsinkable" *Titanic.*

Captain Smith chose to ignore the repeated warning radio messages from different ships in the area of the North Atlantic through which the liner was intending to traverse. Instead, he continued to steam ahead at about 22 knots, trusting on the ability of his six lookouts, two of whom were situated in the crow's nest and exposed to the extreme cold.

When an iceberg was eventually spotted by a lookout, towering up about 15 meters (50 feet) above the oily calm sea, at a distance of just over 457 meters (500 yards), the order of "full astern" and "hard a-port" had little immediate effect on the momentum of the great ship and could not prevent the inevitable disaster.

Even then, many of the crew and most of the passengers did not fully appreciate the magnitude of the impending disaster and were reluctant to enter the lifeboats and abandon a ship they considered unsinkable. Consequently, many lifeboats were lowered into the water without their full complement. Some of these were immediately rowed away toward the lights of a ship that were visible in the distance, thus taking them away from the scene and a position to be able to help others later as the great ship finally sank.

In all, only 712 of the 2,201 persons on board survived. None of the 1,489 people who entered the icy, calm water survived more than an hour, even though most were wearing life jackets. Indeed some survivors later described how the cries for help from those in the water had almost disappeared in half an hour.

Maritime safety regulations of the day, specifying the number of lifeboats to be carried by a ship, were based on the overall tonnage of the vessel rather than the number of people on board. *Titanic,* with her 16 lifeboats and 4 collapsible boats, had an overprovision of lifeboats by those standards. Yet they had capacity for only 1,178 people, 1,023 less than the number of people on board. Regrettably, because of the failure of people to appreciate the urgency of the situation and poor leadership, only 712 people used the boats.[4]

Thus, a series of misjudgments and minor errors led to an appalling tragedy and a needless loss of life. Initial overconfidence in the unsinkability of *Titanic,* reluctance to accept that a problem was developing, and ill preparedness of the crew and passengers for abandoning ship all contributed to the disaster. At several stages in the course of those final hours, a commonsense assessment of the facts available should have resulted in correct decisions and appropriate actions. One of the world's biggest maritime disasters would then have been averted.

---

4. The following statistics, published by *USA Today* (11 January 1999), may be of interest in this respect. "U.S. Coast Guard statistics state that there are about 6,000 registered passenger boats operating in U.S. waters. They carry about 200 million passengers annually (134 million in 35 states). The great majority do not carry 'out of the water' (OTW) craft or safety boats. The nation's largest ferry operator, Washington State Ferries, operates 28 ships with a combined capacity of 37,565 passengers and an OTW capacity for only 8,645 people—less than 25 percent of capacity. When fully laden only 3 ships can accommodate all passengers in OTW craft. Fortunately, to date the safety record is impressive, but accidents are no great respecter of safety records. Most dangerous are probably whale-watching boats that can operate over 19 kilometers (12 miles) offshore in cold waters without OTW safety capability, depending instead on the potential of other craft to come to their rescue in the event of an accident." In an area with low sea temperatures and prone to fog, that is a "brave" decision.

LESSONS LEARNED

From the innumerable lessons that we can learn from the historical accounts of survival at sea, two fundamental lessons appear time and time again. First, the sea has great potential to produce a profound and unexpected challenge to survival. Second, the consequences of this challenge, and the fates of those involved, are often in their own hands. A series of incorrect decisions can result in tragedy. A series of early and correct decisions may result at worst in a heroic story of survival and at best in avoidance of the problem altogether. Recognition of the early danger signs, however, comes only with experience, careful thought, and planning. Those who go to sea should carry out a risk analysis of any activity in a potentially dangerous environment. Such an assessment should make them aware of the potential dangers involved, the protective measures that they should implement, and, more importantly, the critical early warning signs that might indicate impending danger.

Anecdotal evidence clearly indicates that having a survival strategy can make the difference between life and death. The prudent will usually make the necessary arrangements to give themselves a fighting chance of survival. In devising such a strategy it is essential to

- ▶ understand the nature of the threats posed by the environment,
- ▶ understand their effects on the body (physiological effects),
- ▶ assess the immediacy of each threat, and
- ▶ assign priorities based on their chronological sequence.

Plans to counter the threat of entrapment, fire, smoke, being washed overboard, and so on should have a higher priority than those to counter drowning, body cooling, or other immersion problems. Likewise, for those safely out of the water and in life rafts, drinking water has a higher priority than food, but in temperate or arctic climates both are secondary to maintaining body temperature.

Critical to this process of threat assessment is an understanding of how the body normally functions (physiology), how best to maintain this function in adverse conditions, and the level and consequence of functional impairment likely in a survival situation. This book provides critical information in these areas. Such knowledge does increase survival prospects. An understanding that impairment of physical and mental performance can occur rapidly in an adverse environment should lead people to complete certain actions essential to survival early, while doing so is still possible. For example, early impairment of manipulative skills makes the preparation of location aids a high priority, once the immediate threats to life have been safeguarded against.

The cardinal principles on which to plan a survival strategy are

 ▶ protection,
 ▶ location,
 ▶ water, and
 ▶ food.

We examine the bases of these principles in the chapters that follow.

Chapter Summary and Recommendations

 ▶ The sea has the great potential to produce a profound and unexpected challenge to survival.

 ▶ We can learn much from the history of maritime disasters.

 ▶ History repeats itself. Disasters with remarkably similar features continue to occur at sea.

 ▶ Such tragedies frequently occur without warning. Often they are the result of a series of unconnected events (misjudgments, minor errors of omission or commission) that cumulatively lead to a disaster.

 ▶ The speed with which the final events can unfold leaves no time for initial familiarization with, or training in the use of, safety and survival equipment.

 ▶ Thus, the fate of survivors is often in their own hands. There are no substitutes for knowledge, training, and experience.

A good risk analysis is essential in formulating a worthwhile survival strategy. A necessary component of both is an understanding of the nature of the medical-physiological threat likely to be encountered and consequent incapacitation.

Manufacturers of survival and rescue equipment and those who provide training should also understand the medical-physiological threat and make necessary allowance for the consequent impairment of capability.

C H A P T E R

2

Basic Physiology of Survival

"I KEEP SIX HONEST SERVING MEN** (they taught me all I know). Their names are What and Why and When, and How and Where and Who."

Rudyard Kipling, 1865–1936

A basic understanding of how the body works (physiology) is helpful in understanding the what, why, when, and how of survival principles. Without such an understanding, it may be difficult to interpret much of the advice that is given and difficult to refute or confirm the claims of the proponents of some new, "improved" technique or expensive device said to be essential for survival. Just as advances in equipment and clothing technology influence survival strategy, so too should our increasing knowledge of the physiological responses of the body in extreme environments. For example, 40 years ago the accepted wisdom was that it was best to undress when accidentally immersed in cold water. Survival guides gave detailed instructions on how to remove waistcoats, suspenders, and garters in the water! We now know that it is better to remain clothed. Likewise, even today, we find survival books that include details of how to "drown proof" by lying face down in the water with the head submerged except when breathing. We now know that this method will accelerate, rather than delay, death in cold water.

Bearing in mind the old military adage that "the best-made plans frequently disappear once the enemy is engaged," there is a more important reason for understanding the nature of the physiological threat to the body. In many survival situations, essential survival equipment is often damaged or lost; consequently, the survivor must improvise. A good understanding of the nature of the threat and its effect on the body enables one to improvise defensive strategies using whatever materials are at hand. With this philosophy in mind this chapter introduces in general terms the physiological mechanisms relevant to survival at sea. Those pertinent to both water and food intake are dealt with in chapter 8. The important problems associated with body temperature regulation are considered in detail here.

CELL SURVIVAL

The cell is the smallest structural unit into which an organism can be divided and still retain the characteristics we associate with life. The average human is constructed from around 50 billion cells. These cells work in a semiautonomous manner within collective units *(organs)* to ensure the well-being of the body. To survive and function, the cells of the body require an environment that resembles the primeval ocean in which the first cells evolved. In humans the fluid that bathes the cells *(extracellular fluid,* or ECF) provides this "ocean," but unlike the primeval ocean the volume of the ECF is not vast, containing only about 19 liters, or 20 quarts (figure 2.1).

The ECF is the connecting link between the cell and the blood stream, and through the blood stream to the other organs of the body and the outside world. The finite volume of the ECF means that these connections are critical. Through them the chemical constituents of the ECF can be carefully regulated to ensure optimum cellular function. Although the volume and composition of the ECF remain reasonably constant in normal healthy people, one should not assume that it is a static system. On the contrary, dynamic processes operating between the body and the external environment, and internally between the various body-fluid compartments and organs, actively maintain this apparently steady state.

With little room for variation, the body employs many responses and reflexes aimed at maintaining a constant internal environment in the face of a widely varying external environment. The air we breathe, the food and fluids we consume, the exercise we take, all influence the chemistry of the body. The body senses any changes and when necessary initiates a corrective response. A shortage of oxygen in the inspired air triggers increased breathing *(respiration)*; a shortage of water, which causes an increase in the concentration of various salts in body fluid, triggers a desire to drink and reduces urine output; and an overabundance of water in the body causes the kidneys to excrete fluid to maintain the correct concentration of salts in the tissue fluid. Additionally, exercise may produce chemical alterations resulting in an increase in the level of acid in the muscles that would prove harmful to the body if not neutralized

Body fluid

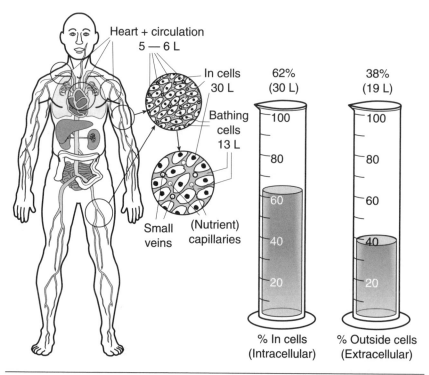

Figure 2.1   The distribution of body fluids. About 65% of body mass is fluid. (1 liter [L] = 1 kilogram [kg], approximately. Thus, 75 kg mass = 49 L of water.)

and removed. The adult human body produces more than 12 liters (12.7 quarts) of industrial-strength acid every day. A biochemical buffering system neutralizes this acid, enabling its excretion through the air we exhale and the urine we produce.

This regulation of the interior environment (termed *homeostasis*), undertaken primarily by the kidneys and respiration, has enabled humans to populate a planet of widely differing environmental extremes.

SURVIVAL BALANCE

Humans belong to a group of animals called *homeotherms*. To ensure optimal physiological function and survival, these animals must regulate their deep body temperature within a narrow range despite large changes in environmental temperatures. To do this, their heat loss and heat production must be in balance. Indeed, many will be familiar with the phrase "his survival was in the balance," and in one sense this neatly summarizes the fundamental basis of survival.

To survive, the body, through homeostasis, must remain in balance, not just one balance, but several. To avoid an increase or fall in deep body temperature, we must achieve thermal balance (figure 2.2); to avoid dehydration, we must achieve water balance; and to avoid starvation, we must remain in energy balance. Both unconscious (automatic responses) and conscious responses (driven by hunger, thirst, and thermal discomfort) work to maintain these balances. We will return to these topics in later chapters when we discuss the physiological requirements relating to both food and water. In the remainder of this chapter we focus on heat balance, a balance that determines body temperature and is therefore critical to survival at sea.

The chemical reactions occurring within the cell, which keep it alive and functioning, produce heat. This heat is a by-product of the work of living cells, termed *metabolism*. The greater the activity, the greater the amount of heat produced. This heat, or thermal energy, may be quantified in terms of the temperature of the body. For this purpose we can divide the body, somewhat simplistically, into an irreducible *core* of vital organs and a variable *shell* of superficial tissues. Core temperature is controlled by increasing (in the cold) or decreasing (in the heat) the thickness of the shell in ways described later in this chapter.

Metabolism is measured in watts, and in the resting state the average human produces about 100 watts (about the same as a domestic light bulb) through the normal cellular activity involved in staying alive. Exercise produces additional heat, primarily from the muscles, up to a maximum of about 1,500 watts (1.5 kilowatts), equivalent to a domestic electric heater.

Cells are influenced by temperature, and the temperature of a cell is determined by its energy turnover and by the rate with which heat is removed from it. Optimum function in humans occurs at a temperature that is around the normal body temperature of 37 degrees Celsius (98.6 degrees Fahrenheit). As body temperature falls, brain cell activity, and therefore function, slows and eventually ceases at around 17 degrees Celsius (63 degrees Fahrenheit). Before deep body temperature reaches that level, however, cooling of the extremities will occur. Even moderate cooling of the extremities will impair function, and more profound cooling will result in injury to the skin, muscle, and bone of the exposed part (*cold injury*; see chapter 10).

In engineering terms, muscles are about as inefficient as the internal combustion engine in a family car, and use only 25 percent of the energy they consume to perform work, releasing the remaining 75 percent as a

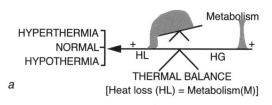

Thermoneutral Environment

a

THERMAL BALANCE
[Heat loss (HL) = Metabolism(M)]

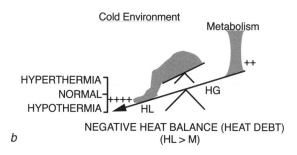

Cold Environment

b

NEGATIVE HEAT BALANCE (HEAT DEBT)
(HL > M)

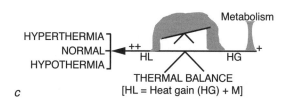

c

THERMAL BALANCE
[HL = Heat gain (HG) + M]

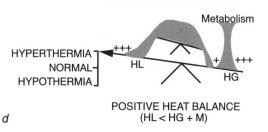

Hot Environment & Excercise

d

POSITIVE HEAT BALANCE
(HL < HG + M)

Figure 2.2 Diagrammatic representation of heat balance in different environmental conditions. *(a)* In a thermoneutral environment, heat loss counters heat produced by metabolism to achieve thermal balance. *(b)* In a cold environment, when body heat loss exceeds that produced by metabolism, the result is a fall of body temperature. *(c)* In a hot environment, heat loss increases to counter heat gained from the environment and produced by metabolism; again, thermal balance is maintained. *(d)* When exercising, especially in a hot environment, should heat loss be unable to keep pace with heat produced, the result will be a rise in body temperature.

by-product, heat. This heat must be removed from the body to prevent an undesirable rise in body temperature. For example, most individuals will be unable to tolerate a deep body temperature of 42 degrees Celsius (107.6 degrees Fahrenheit) with a skin temperature of 38 degrees Celsius (100 degrees Fahrenheit) without suffering heat illness. These body temperatures represent an increase in the amount of heat stored in the body of a 70-kilogram (154-pound) human of about 250 kilocalories.[1] If the body were unable to lose any of the heat it produced, it would reach a fatal level of heat storage in about four hours when at rest and after just 25 minutes with moderate exercise.

Thus, heat gain can become critically important, even in a temperate climate, for those exercising in clothing that significantly impedes heat loss. Such clothing includes that used by firefighters or those dealing with chemical spills. This excessive heat load is the explanation for several fatalities that have occurred in such scenarios. In the year 2000, 112 firefighters lost their lives in the United States. The single largest cause of death (46) was cardiac problems related to overheating *(hyperthermia),* caused by the inability to lose metabolic heat. At tissue temperatures of about 45 degrees Celsius (113 degrees Fahrenheit), the protein in cells is altered and can be irreversibly damaged, just as the protein in eggs will coagulate as they are heated. This effect has the greatest consequence in brain tissue. Before the body reaches lethal temperatures, function is impaired (see chapters 6 and 10).

Thermal (Heat) Balance: Basic Physics

To remain in thermal balance, the body must transport the heat that it produces to its surface and lose it to the environment. The first two laws of thermodynamics can describe heat exchange between an object and the environment.[2] Heat flows down a thermal gradient (that is, from high to low temperatures). Thus, when air temperature is cooler than body temperature, cooling is primarily achieved by the circulating blood, which is heated as it passes through the deeper tissues of the body (core) and cooled as it passes through the blood vessels of the skin (shell). Heat also flows through the tissues of the body (by conduction—see follow-

---

1. One kilocalorie is the amount of heat required to raise the temperature of one liter of water by one degree Celsius. 1 watt (W) = 1 joule per second = 0.86 kilocalories per hour.

2. The three laws of thermodynamics can be stated as follows: (1) energy cannot be created or destroyed but is transformed from one form to another (conservation of energy); (2) heat flows down a thermal gradient (that is, from high to low temperatures); and (3) the temperature of a system cannot reach absolute zero (–273 degrees Celsius) in a finite number of operations.

ing) from the warmer, deeper tissues to the cooler tissues near the surface of the body. For heat to escape from the body it must be transferred from the skin to the environment. In normal circumstances in air, the body can exchange heat with the environment by four physical processes: radiation **(R),** convection **(C),** conduction **(K),** and evaporation **(E)** (figure 2.3). For body temperature to remain stable in a cool environment (below body temperature), metabolism **(M)** must produce enough heat to match that lost by R, C, K, and E. In a hot environment (above body temperature), heat produced by metabolism (M) and that gained by R, C, and K must be matched by that lost by E. Heat exchange between the body and its surrounding environment, and within the body between tissues at different temperatures, is best illustrated by the general heat-balance equation

$$M - W = E \pm C \pm K \pm R \pm S$$

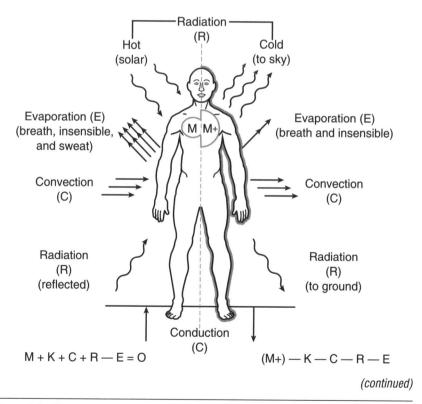

(continued)

Figure 2.3   Diagrammatic representation of the routes of body heat gain and loss in a naked person in air and water.

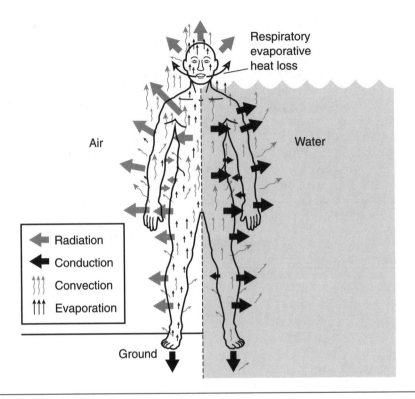

Figure 2.3 *(continued)*

where the unit for each term is watts per square meter (W/m²) and

- ▸ **M** is the total metabolic rate of the individual,
- ▸ **W** is measurable external work, and
- ▸ **S** is heat storage.

If thermal balance is maintained, the terms of the equation should sum to zero and the body temperature should remain stable (i.e., S = 0).

R *(Radiation).* All objects possessing heat emit thermal radiation from their surfaces in the form of a wave of energy containing particles (photons) within the red-infrared range of the electromagnetic spectrum. The energy from these particles is absorbed by, and transferred to, the atoms of objects they come into contact with. No medium is required for the transfer of heat by radiation, and the temperature of any air through which heat radiates has little effect on the heat transferred. Thus, radiation is the process by which the energy of the sun travels through the vacuum of space to reach earth and power life on it. Radiation is also the process that enables the body to gain heat from the sun on a cold day and lose it to a cloudless sky at night in the desert. Radiation can be a major route of heat loss in certain scenarios, for example, in poorly clothed individuals in an open boat on a clear, cool night. Heat loss by radiation is negligible in water. The quantity of heat transferred from one object to another by radiation depends on the

▶ effective radiating surface areas,

▶ difference between the mean surface temperatures of both objects, and

▶ emissivity (proportion of maximum radiation emitted) and reflectivity of an object in relation to the emissivity and average radiant temperature of the environment.

When a person stands with arms and legs extended away from the body, the radiating surface area is maximal at about 85 percent of total body surface area. When a person stands with legs together but arms extended, the radiating surface area is about 75 percent of the total. When the body curls up in the fetal position, the area is about 50 percent of the total.

C *(Convection)*. The body of a naked person standing in cool air (below skin temperature) will warm the air molecules that contact it. Consequently, the density of this air will decrease, and the air will rise, to be replaced by cooler molecules. This process is called *natural (free) convection*. (See also the later section "Boundary Layer.") Air (wind) or water (current) movement across the skin (called *forced convection*) or the movement of the body in air or water (relative wind speed or relative water current) increases convective heat exchange. Thus, the exchange of heat between a body and its environment through convection depends on the

▶ temperature gradient between the two (this determines the amount of heat absorbed or donated by a given mass of air that comes into contact with the skin), and

▶ relative movement of the fluid (air or water) in which the body is placed.

K *(Conduction)*. This term describes heat exchange between the skin and surrounding surfaces with which it is in direct contact. With this form of energy exchange no physical movement of material occurs within the respective objects. Rather, vibrating molecules exchange energy directly between themselves. Usually the amount of heat exchanged in this way is small. For an upright naked person in air, the only surface area available for conductive heat exchange is that of the soles of the feet and the ground on which the person stands. This area is a small percentage of the surface area of the body and a small percentage of the total heat exchange with the environment. The amount of heat exchanged by conduction depends on the

- ▶ temperature gradient between the skin and the surface with which it is in contact,
- ▶ surface area in contact, and
- ▶ thermal conductivity (ease with which heat moves through a substance) of the surface in contact with the skin.

From this it is clear that in some circumstances conduction can become a major route of heat exchange. An individual who lies down increases the surface area of the body in contact with the ground and, depending on its temperature, can experience significantly greater conductive heat exchange. Water has about 24 times the thermal conductivity of air, so heat loss increases significantly on immersion in cold water.

Thermal exchange by R, C, and K are variously referred to as *dry heat transfer, sensible* (measurable) *heat transfer,* or *nonevaporative heat transfer,* terms that distinguish these routes from evaporative heat transfer.

E *(Evaporation).* Evaporation is the process by which energy transforms liquid to a gas. Heat is transferred from a warm body to water molecules on its surface. With the addition of thermal energy, the temperature of this water rises and evaporation occurs. Thus, evaporative heat exchange occurs only within a gaseous medium and when water evaporates from the surface of an object. The heat required to drive this process is removed from that surface and is cooled. This is termed the *latent heat of vaporization,* and for water it amounts to 576 kilocalories per liter (1.05 quarts). Even in a cold environment the body constantly loses water by evaporation from the lungs and skin to the surrounding environment. For example, on a cold day the moisture vapor in expired air is readily visible as a fog; so too is the condensation that forms on the inner surface of waterproof clothing. This *insensible* water loss, which is outside the body's control, amounts to about one liter per day (chapter 8) and, if evaporated, will dissipate about 24 kilocalories per hour (21 watts).

The rate of evaporation depends on the

- ▶ skin surface area that is wet,
- ▶ air movement around the body (wind or body movement), and
- ▶ difference between the *vapor pressure* at the skin surface and that in \the air. The vapor pressure on the skin is proportional to skin temperature; in the air it depends on air temperature and relative humidity.

Boundary Layer

The term *boundary layer* describes the layer of molecules of air or water immediately adjacent to the skin. These molecules move more slowly and are less disturbed by natural convective currents. Consequently, they are stiller than those a few millimeters distant from the surface. Over naked, hairless skin, the thickness of this boundary layer is a few millimeters; it increases if hair is present. The thicker the layer, the better its insulative value. Animals and birds make use of this fact by increasing the amount of air in the boundary layer by involuntarily raising their hair or puffing up their feathers in cold weather. "Goose bumps" are a vestigial remnant of this mechanism in humans.

The molecules in the boundary layer are heated to skin temperature by conduction and act as a thermal buffer against the warmer or colder molecules in the air or water immediately outside the layer. Although the innermost molecules in the boundary layer remain relatively still, natural convective currents will influence those in the outer reaches. This small mass of relatively still molecules in the boundary layer plays an important role in insulating the skin of naked humans in both cold and hot environments. Thus in still air on a cool day, a naked person may happily remain in thermal comfort until a gentle breeze disturbs the boundary layer. Conversely, in a hot, humid environment such as a sauna, a fan can cause a tolerable temperature to become uncomfortable. In cold or hot water, gentle movement can quickly disturb thermal comfort.

In air, the stronger the wind, the thinner the boundary layer will become. Even low air speeds easily disturb the outer molecules of the boundary layer. Therefore, a small increase in wind speed in relatively calm conditions produces a much greater reduction in the thickness of the boundary layer than the same increase in speed when the wind is already strong. Increasingly faster air speed eventually destroys the boundary layer, and further increase in wind speed (or water current) has little additional cooling effect. So it is that increasing air movement above 15 kilometers per hour (9.3 miles per hour) or water flow above 0.5 meters per second (1.8 kilometers per hour, or 1.1 miles per hour) has little additional heating or cooling effect on the body. Swimming at any speed maximizes convective heat loss.

Application to the Body

Although it is useful to describe the routes of heat exchange separately, in reality they occur together. Thus, the metabolic heat of a clothed person in a cold environment is conducted from the skin to the clothing in contact with it. It moves by convection within the clothing and is conducted to the surface of the clothing, where it is lost to the environment by convection, radiation, and evaporation.

In water only two of the four primary pathways for heat exchange are available. Heat is conducted to the molecules of water in the boundary layer in contact with the skin (see figure 2.3). These warmed molecules rise because of convective movement, and cooler ones replace them. Thus, in water, heat loss is principally by convective and conductive heat exchange. Despite this, a naked individual in cold water will cool approximately four times faster than when in air at the same temperature. Water can produce a physiological response equivalent to that produced by air 11 degrees Celsius (20 degrees Fahrenheit) cooler. This disparity occurs because the thermal conductivity of water is 24 times that of air and its volume-specific heat capacity[3] is approximately 3,500 times that of air (table 2.1). Therefore, water has a much greater capacity to extract heat. Furthermore, a body in water, unlike a body in air, presents a surface area for heat exchange with the environment that is close to 100 percent.

Table 2.1   Physical Characteristics of Air and Water

	Thermal conductivity mW/m²/°Kelvin	Specific heat J/g/°Kelvin	Density g/cm³
Air	26.2	1.007	0.0012
Water	630.5	4.1785	0.9922

PHYSIOLOGICAL REGULATION (THERMOREGULATION)

A relatively small number of species, including most birds and mammals, control their deep body temperature. They do this by both altering their behavior (see chapter 3) and initiating physiological responses.

In humans it is the role of the *thermoregulatory system* to vary the level of heat exchange with the environment to maintain deep body temperature within about half a degree Celsius (1 degree Fahrenheit) of the normal deep body temperature of 37 degrees Celsius (98.6 degrees Fahrenheit). The body defends its temperature with some vigor, and the thermoregulatory system can take precedence over other regulatory systems.

---

3. The volume-specific heat capacity is obtained by multiplying the specific heat of a substance by its density. It represents the amount of heat required to raise the temperature of a given volume of water by 1 degree Kelvin. At 37 degrees Celsius (98.6 degrees Fahrenheit) the volume-specific heat capacity of water is 3,431 times that of air.

For example, the body maintains its temperature during starvation when it could save energy by permitting body temperature to fall; blood-pressure regulation can fail because of increased circulation to the skin *(vasodilatation)* in the heat; and the body continues to sweat when depleted of water *(dehydration)* although cessation of sweating would conserve water.

Thermoneutral temperature is defined as the environmental temperature at which the body can maintain normal deep body temperature simply by altering the amount of blood, and therefore heat, that flows to the skin, and without recourse to shivering or sweating. For a naked person, thermoneutral air temperature is 26 to 30 degrees Celsius (79 to 86 degrees Fahrenheit) in still air and 35 to 35.5 degrees Celsius (95 to 96 degrees Fahrenheit) in water. This range has been given various names including the *vasomotor zone,* the *intrathreshold zone,* and the *null zone* (figure 2.4). These terms refer to the fact that within the thermoneutral zone, no shivering or thermoregulatory sweating occurs because neither is required to maintain body temperature. The gap between shivering and sweating created by the ability of the body to thermoregulate by changing the amount of blood flowing to the skin is important because it stops the body from repeatedly switching from sweating to shivering. To do so would be uncomfortable and wasteful of energy reserves (used in shivering) and water (used in sweating).

When in the thermoneutral zone, the average skin temperature is about 33 degrees Celsius (91 degrees Fahrenheit). This high temperature is consistent with a species that originated in a hot environment. The fact that humans are relatively hairless and begin to defend their deep body temperature when air temperature falls below 26 degrees Celsius (79 degrees Fahrenheit) lends further support to this hypothesis. Furthermore, humans have more sweat glands than any other mammal and can sweat at a high rate, suggesting that the warm environment in which they evolved had low humidity and thus enabled evaporative cooling. As environmental temperature increases or decreases, a point is crossed where the body starts to sweat or shiver. The body-temperature thresholds at which these responses occur, and the rate at which the responses increase with rising or falling body temperature, can be used to characterize the thermoregulatory system. These thresholds and rates vary between people and, as we shall see later, with a range of other factors.

Once initiated, shivering or sweating help maintain body temperature over a range of temperatures called the *thermoregulatory zone*—the range of environmental temperatures within which a person is able to thermoregulate using their physiological thermoregulatory responses. Outside the thermoregulatory zone a naked person will be unable to thermoregulate. As a consequence, body temperature will continue to rise (or

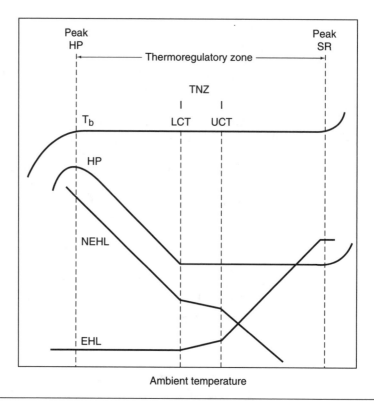

Ambient temperature

Figure 2.4   Thermal thresholds and zones. Peak HP = peak heat production; peak SR = peak sweat rate; thermoregulatory zone = range of ambient temperature in which the naked body can thermoregulate; TNZ = thermoneutral zone, ambient temperature range in which the body can thermoregulate through vasomotor action alone; T_b = deep body temperature; LCT = low critical threshold (that is, body temperature at which shivering commences); UCT = upper critical threshold (that is, body temperature at which sweating commences); HP = heat produced; NEHL = nonevaporative heat loss; EHL = evaporative heat loss.

Adapted from Mekjavic and Bligh 1987.

fall) to the point of death unless other strategies such as external heat sources, clothing, buildings, air conditioning, and so on are used. The behavioral or technological zone represents the range of environmental temperatures in which humans can thermoregulate if they have the appropriate external resources and facilities. This zone can cover all climatic temperatures encountered on earth. Therefore, to meet the principal challenge to survival, an individual must have the necessary equipment and skills to enable his or her body to thermoregulate in its environment. The principles of behavioral thermoregulation as they relate to clothing are discussed in the next chapter.

Wearing dark clothing to maximize heat gain from solar radiation, as these Merchant Navy sailors are doing, is one strategy to keep body temperature from falling.

Photograph courtesy of National Archives of Canada.

Thermoregulatory System

To control body temperature, the thermoregulatory system must be able to sense temperature and respond to changes. Thus, it includes receptors that respond to rises *(warm receptors)* and falls *(cold receptors)* in temperature. These receptors are located in areas of the body such as the skin, muscles, spinal cord, and brain. In the skin, for example, there are about three times as many cold receptors as there are warm receptors. Because they are located about 0.18 millimeters (one 100th of an inch) below the surface of the skin, they respond quickly to changes in environmental temperature. In contrast, the area of the brain that is particularly sensitive to temperature, called the *hypothalamus*, contains more warm than cold receptors. These are not stimulated until changes in environmental temperature or metabolic heat production (e.g., exercise) affect the temperature of the brain by the flow of blood through it. With regard to the initiation of thermoregulatory responses in a human, the temperature receptors in the brain are much more sensitive to temperature change than those in the skin.

Besides being sensitive to its own temperature, the hypothalamus receives and processes thermal information from the remainder of the body and initiates appropriate responses.

Response to Falling Temperature: Increased Heat Conservation

When its temperature starts to drop, the body begins conserving heat through two methods: altering the flow of peripheral blood and raising the hairs on the skin.

Alteration in peripheral blood flow.

When cold, the body shuts (*vasoconstricts*) the blood vessels in the skin (figure 2.5), reducing blood flow to the skin and the underlying fatty layer. In so doing, it converts both the skin and the fatty layer into an insulating buffer zone on the surface of the body that protects the inner core temperature where the vital organs are situated. Insulation is defined as resistance to heat flow. The thicker the layer of fat beneath the skin (*subcutaneous fat*), the better the resulting insulation in the cold. Thus, reducing blood flow to the skin is the means by which the body employs peripheral insulation. The fat beneath the skin can be regarded as fixed insulation that changes little in the medium term. It provides approximately the same insulation as cork (1.5 clo per centimeter, or about 4 clo per inch, of fat; see chapter 3). Muscle can also provide significant insulation when it is relatively unperfused by blood (that is, when at rest). This source of insulation, however, is lost at relatively low levels of exercise, including shivering ("variable" insulation).

The reduction in skin blood flow decreases the amount of heat delivered to the surface of the body from the core. Consequently, skin temperature decreases and becomes closer to the temperature of the environment, reducing the gradient down which heat can be lost to that environment.

The ability of the body to change skin blood flow is great. As an example, total skin blood flow when the blood vessels open maximally (*vasodilated*) in the heat is 3 to 4 liters (3.2 to 4.2 quarts) per minute. This is reduced by 99 percent, to about 0.02 liters per minute, when maximally vasoconstricted in the cold. We can identify at least three functionally different regions with regard to the control of skin blood flow: the extremities (hands, feet, ears, lips, nose); the trunk and upper limbs; and the head and brow. Although the body uses most of its outer layer (the shell) as a variable insulator, the one exception is the scalp, where blood flow tends to remain constant regardless of temperature changes. In cold environments about 50 percent of total body heat production may be lost through the unprotected head of a lightly clothed individual in an equivalent air temperature of –4 degrees Celsius (24.8 degrees

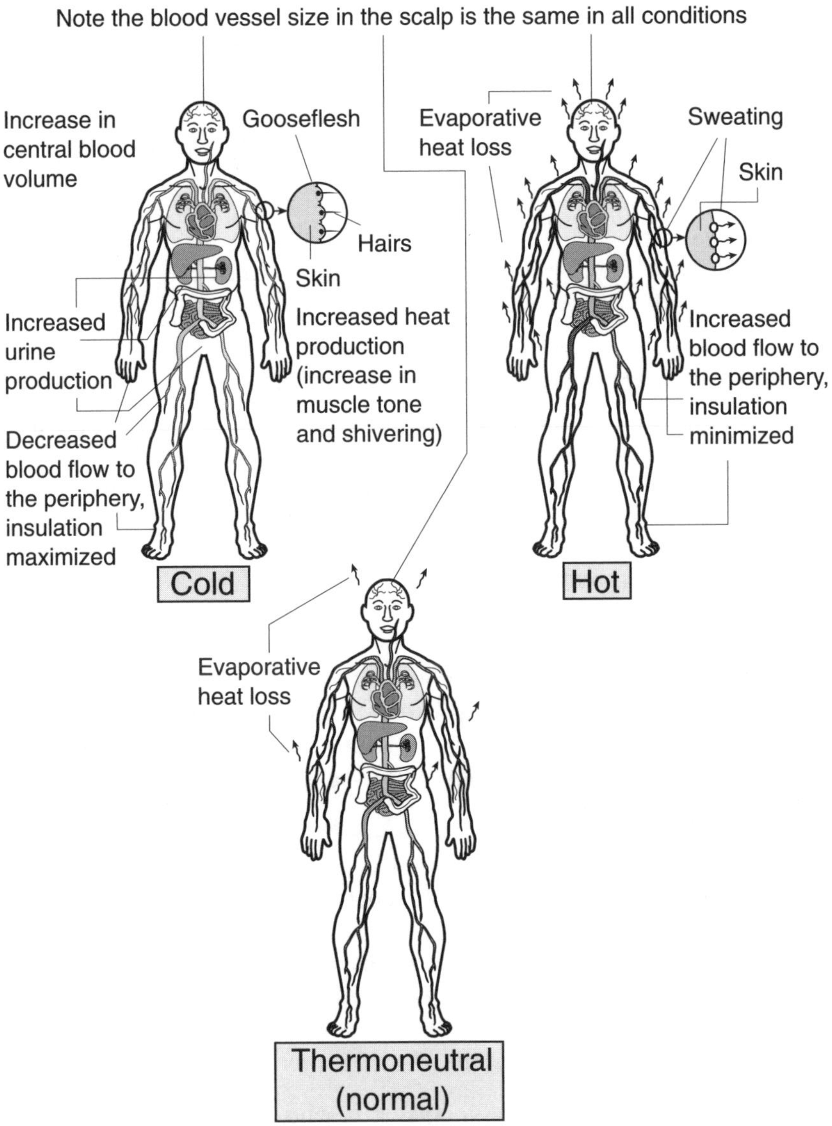

Note the blood vessel size in the scalp is the same in all conditions

Increase in central blood volume

Gooseflesh

Evaporative heat loss

Sweating

Skin

Hairs

Skin

Increased urine production

Increased heat production (increase in muscle tone and shivering)

Increased blood flow to the periphery, insulation minimized

Decreased blood flow to the periphery, insulation maximized

Cold

Hot

Evaporative heat loss

Thermoneutral (normal)

Figure 2.5 Diagrammatic representation of the effect of skin blood flow on heat exchange.

Fahrenheit).[4] When more clothing is worn a relatively higher percentage of total heat loss will be lost through the unprotected head.

In cold environmental conditions, the vasoconstriction of superficial blood vessels shunts blood away from the periphery and temporarily increases the circulating blood volume in the center of the body. Nerve (stretch) receptors in the large blood vessels near the heart immediately sense this overload of the central circulation. Appropriate redistribution measures are reflexly initiated, including an increase in the production of urine *(cold diuresis)*. This is a familiar phenomenon to most people when they enter a cold environment, and particularly on cold-water immersion.

Raising the hairs on the skin (*piloerection,* or "goose bumps").

Most animals use this technique to conserve body heat in cold environments. For the same reason, some animals from cold climates have thick coats of fur with different hair lengths, some of which are hollow, to enable even more air to be trapped next to the skin (increase in boundary layer). Because humans evolved in warm climates they are relatively hairless and thus, although piloerection does occur, it provides little insulation.

Response to Falling Temperature: Increased Heat Production

In addition to conserving heat through the methods mentioned above, the body also reacts to a drop in temperature by increasing its heat production. It does this through increasing muscle tone and shivering.

Increase in muscle tone.

As the body cools, an increase in muscle activity (tone) occurs that is not associated with any movement. This is part of the explanation for the stiffness one feels when cold. Because this increase in heat production occurs immediately, before the thermal gradient for heat loss from the deep body tissues has been established, deep body temperature may increase a little on initial exposure to the cold (see chapter 6).

Shivering.

Shivering is the involuntary, synchronous, and rhythmic contraction of small parts of skeletal muscles called *motor units*. These contract at a rate of about 10 to 20 per second, out of phase with other units. The contractions alternate with motor units of opposing muscles so that large

---

4. Equivalent air temperature, or more properly, equivalent still-air temperature, is the term used to indicate that wind chill should be taken into account. That is, the cooling effect of wind speed at higher ambient temperatures will result in the same rate of heat loss as would occur at a lower temperature in still air. See the section "Wind Chill" in chapter 10.

movements do not occur and no external work is done. Heavy shivering may be interspersed with periods of light shivering or rest in the early phase of cooling, but shivering later becomes continuous before progressing into an almost tonic state. At its maximum, heat production from shivering can reach five or six times the amount that occurs at rest. But the body cannot maintain this level of shivering heat production for long.

Because no external work is performed, all the energy liberated by shivering appears as heat. Some of this heat is immediately lost to the environment, but in most situations shivering can be an effective, if uncomfortable, way to maintain body temperature.

The energy needed for shivering comes from fats and sugar. Of the two, sugar (carbohydrate) is in the shortest supply, and when it runs out shivering stops (chapter 8). Even at moderate levels, shivering can cease in as little as seven hours when an individual eats no food. Shivering also diminishes when oxygen levels in the inspired air fall or when carbon-dioxide levels increase. This factor becomes important in situations where ventilation may be inadequate, such as in life rafts (chapter 9). Shivering uses the same skeletal muscles as voluntary exercise, and the two can coexist up to moderate levels of voluntary activity. With mild cooling, shivering is progressively inhibited as exercise intensity increases. With severe cooling the increase in muscle tone associated with shivering can inhibit coordinated movement and impair activities such as swimming (chapter 4).

In cold climates, should the body be unable to produce enough heat to balance that being lost to the environment, body temperature will inexorably decline and death from hypothermia will occur when deep body temperature has fallen by about 12 degrees Celsius or 22 degrees Fahrenheit (chapter 6).

Response to Rising Temperature: Increased Heat Loss

When the body begins to experience an increase in temperature, it alters peripheral blood flow and sweats in order to rid itself of the excess heat.

Alteration in peripheral blood flow.

When warm, the body vasodilates the blood vessels in the skin (see figure 2.5). This process not only bypasses the insulation of the subcutaneous fat but also increases skin blood flow and thereby the amount of body heat delivered directly to the surface of the body. Consequently, skin temperature rises and, when environmental temperature is lower than body temperature, the gradient down which heat can be lost becomes steeper. For each liter (1.05 quarts) of blood at 37 degrees Celsius (98.6 degrees Fahrenheit) that flows through the skin and returns to the deeper tissues at 36 degrees Celsius (97 degrees Fahrenheit), the body loses roughly one kilocalorie of heat. Because the maximum skin blood flow can be as much as 3 to 4 liters (3.2 to 4.2 quarts) per minute, the same 1-degree Celsius (1.6-degree Fahrenheit) fall in the temperature of the blood would off-load up to four kilocalories per minute (280 watts). Deep body and average skin temperature have a greater effect on peripheral blood flow than local temperature. For example, hand blood flow will remain high if deep body temperature is raised, even if the hand is placed in cold water. Alternatively, it is difficult to increase hand blood flow by heating the hand of a hypothermic individual. The next chapter discusses the implications of this for protecting the extremities.

Exercise not only increases the thermal load on the body but also places an additional burden on circulation. In a hot environment this may become problematic. In cool conditions skin blood flow during exercise approximates 600 milliliters per minute (2 percent of cardiac output). In the heat this can increase to 3 to 4 liters per minute (approximately 15 percent of cardiac output). This large amount of skin blood flow not only reduces the blood available for the muscles during exercise but also can increase the demand on the heart and sometimes exceed its capability. This problem becomes more acute in individuals who are dehydrated (whose circulating blood volume may already be depleted) or in those with impaired cardiac performance (e.g., the unfit or those with heart disease). If a person is consuming insufficient fluid to replace the water lost through sweating, the additional circulatory demands will cause precious intracellular fluid to move into the circulation at the expense of cell function.

Sweating.

In contrast to most other mammals that have none, humans have about 2.5 million sweat glands. These are distributed fairly evenly over most of the body surface in densities ranging from 100 to 600 per square centimeter (650 to 4,000 per square inch). Although the sweat produced by these glands is 99 percent water and the most dilute of all body secretions, it does contain some salt and is acidic. Consequently, two disadvantages of sweating are loss of salt and water. People acclimatized to heat produce a greater volume of sweat, but its salt content is less than in those who are unacclimatized. This response helps those acclimatized to heat to conserve salt.

Like other thermoregulatory responses, sweating is evoked by a combination of skin and deep body thermoreceptors. Thus skin cooling can inhibit sweating even when deep body temperature is high. With increasing thermal stimulation, sweat appears on the extremities and proceeds centrally. Progressive recruitment of more glands thus increases sweat output. In addition, the body can increase the amount of sweat produced from each sweat gland. As environmental temperature rises above body temperature and the body gains heat by other routes of heat exchange, evaporation of sweat becomes the only route of heat loss from the body.

Evaporation of sweat cools the skin and the blood flowing through it. The cooler blood, leaving the skin in the veins, returns to deep circulation where it can pick up more heat. This process will occur if the skin is wetted by any liquid that evaporates. "Artificial" sweat can be useful in a hot environment when body-water conservation is required, for example, in a life raft.

Humans can produce up to 2 liters (2.1 quarts) of sweat per hour when fully hydrated. But because sweat evaporation, not sweat production, cools the body, sweat that drips off the body has no cooling effect. It results merely in dehydration. Each liter of water (sweat) evaporated removes 576 kilocalories of heat. For a given air temperature, the rate of sweat evaporation falls as humidity increases. Therefore, it is easier to maintain body temperature and feel cooler in a hot, dry environment (high rate of sweat evaporation) than in a hot, humid environment (low rate of sweat evaporation), even though air temperature and sweat production may be the same in both. Those who have experienced the immediate effects of adding water to the heating source in a sauna readily understand this phenomena.

The Cooling Power of Sweating

In his historic experiment of 1775, Blagden provided an interesting example of the importance of the cooling power of sweating in hot, dry environments. Men (able to sweat) remained in an oven for a period of 15 minutes in a dry atmosphere of 119 degrees Celsius (246 degrees Fahrenheit) without suffering ill effects, while a beefsteak (unable to sweat) on an adjacent shelf fully cooked! Both the steak and the men would have "cooked" had the oven been hot and humid rather than hot and dry.

In hot, humid climates when the thermal gradient is from the environment to the skin, the body will gain heat, making it difficult to control body temperature, particularly during exercise. In such climates, body temperature can continue to rise, leading eventually to *hyperthermia*, heat illness, heat stroke, and death when body temperature has risen by about 8 degrees Celsius (14 degrees Fahrenheit) (see chapter 10).

Nonthermal Factors Influencing Thermoregulation

The preceding discussion makes it clear that thermal factors such as skin and deep body temperature influence the thermoregulatory system. But a wide range of nonthermal factors also affects it (table 2.2). Some of these alter the way in which thermoreceptors in the hypothalamus respond to changes in temperature. Others, such as fitness, alter the body's ability to produce heat or sweat. Each can alter the onset of sweating and shivering, the sensitivity of those responses, and thus the width of the vasomotor zone (see figure 2.4 on page 30).

Table 2.2   Nonthermal Factors That Influence the Thermoregulatory System

Age	Intoxication (drugs or alcohol)
Gender	Hypoglycemia (low blood sugar)
Fitness level	Adaptation to heat or cold
Illness	Raised ambient carbon dioxide
Injury	Lowered ambient oxgen

Chapter Summary and Recommendations

▶ An understanding of how the body responds (physiological responses) to environmental stresses should underpin most survival teaching.

▶ Such knowledge should enable you to conduct a worthwhile risk assessment and formulate a survival strategy for a variety of environments. Understanding body responses will also help you evaluate potential survival aids and improvise with materials at hand should some aids become damaged or lost.

▶ In normal circumstances the body maintains a series of critical balances, including body heat loss or body heat production, water loss or water conservation, and energy expenditure or energy intake. When these balances fail, possibly because of overwhelming environmental challenge, death may ensue. This chapter concentrates on the important question of thermal balance. Chapter 8 covers water and energy balance.

▶ To maintain normal body function, you must tightly control the internal environment in which body cells exist. Little chemical or thermal variation is possible. The dynamic processes that maintain the internal environment are the product of many physiological responses and reflexes. Those relating to thermoregulation include vasoconstriction, vasodilatation, sweating, and shivering.

▶ Radiation, convection, conduction, and evaporation can exchange heat between the body and the environment. The thermoregulatory system attempts to balance this exchange and body heat production to maintain deep body temperature.

3

Behavioral Thermoregulation

"THE BULK CARRIER *FLARE* BROKE UP off Newfoundland in January 1998. The water temperature was 3 degrees Celsius (37 degrees Fahrenheit) and the air temperature 4–5 degrees Celsius (39–41 degrees Fahrenheit). I was lowered down from the helicopter to find six people. They had managed to launch a lifeboat, which had capsized and was floating inverted. One man had slipped off and was dead in the water; a second man was also dead, secured to the boat by a rope. The other four men had been on top of the lifeboat for 5–6 hours, saturated by sea spray. They were scantily clothed and in poor condition, but one person had realised the potential danger and put on every item of clothing that he could find before abandonment. He was coherent and in excellent shape."

Sgt. Tony Isaacs, Canadian SAR

The ability of humans to inhabit and survive extreme climates and to populate and dominate the rest of the planet has been dependent upon intellect and the consequent ability to make use of fire, clothing, and shelter. As the incident just outlined demonstrates, the same applies in a survival situation. Appropriate behavior, based on an understanding of how the body works and what it requires, can make the difference between life and death. We have seen in the preceding chapter how physiological thermoregulatory responses help maintain the body temperature of a naked person for a short period in environmental temperatures ranging from −10 to 150 degrees Celsius (14 to 302 degrees Fahrenheit). These responses, however, are stressful and wasteful of energy (shivering) and body water (sweating) and are therefore finite. As one would expect from a tropical animal, prolonged survival is possible only in relatively high environmental temperatures ranging from 26 to 30 degrees Celsius (79 to 86 degrees Fahrenheit).

Behavioral thermoregulatory responses are conscious activities undertaken to alleviate thermal stress. Although not peculiar to humans, it is they and their ancestors who have been the most effective at employing them to maintain thermal balance. The oldest building discovered has been identified as a windbreak, built by Australopithecus in southern Africa 3.25 million years ago. Clothed bodies dating back 35,000 years have been discovered, and even Neanderthal man was probably using pelts for insulation. Modern humans had populated most of the planet by 10,000 years ago.

Technological progress has brought air-conditioning, central heating, vapor-permeable clothing, and variable-insulation clothing. Yet if we measure the average skin temperature of a heavily clothed Lapp living in the Arctic, a lightly clothed office worker in New York, or a man walking on the moon, we'll find they will not only be about the same as each other, around 33 degrees Celsius (91 degrees Fahrenheit), but also approximate the average skin temperature of the earliest humans living on the east African savannah. Thus, our evolution, and in particular our technological development, has principally enabled us to change *where* rather than *how* we thermoregulate.

Through building, we are able to protect against the extremes of nature by creating a *macroenvironment* (inside a building), the temperature of which we can control. By wearing clothes we can create a tropical *microenvironment* next to the skin that, depending on the number and type of clothes we wear, we can also control. Behavioral thermoregulatory responses, driven by thermal comfort, are clearly much more powerful than our physiological responses. They enable us to survive indefinitely in conditions in which we would perish in a short time if we only had our physiological responses to rely on. In cold climates, humans have adapted behaviorally through the use of shelter, fire, and clothing. In hot climates, behavioral adaptation includes the judicious use of shade, the adoption of appropriate work-rest cycles, and, more recently, air-conditioning.

Behavioral thermoregulation plays an important role in survival at sea. We can regard the acquisition of appropriate lifesaving equipment, such as a life jacket, immersion suit, or life raft, as a form of behavioral thermoregulation. Likewise, we can consider improvisation as behavioral thermoregulation because it optimizes the use of other materials and items to enhance survival prospects.

PRINCIPLES OF CLOTHING
IN RELATION TO TEMPERATURE REGULATION AND SURVIVAL

"Probably every new and eagerly expected garment ever put on since clothes came in, fell a trifle short of the wearer's expectations."

Charles Dickens, Great Expectations (1861)

To have an expectation of the performance of a protective garment, a person must appreciate the requirements of the body in a specific environment and have an idea of the extent to which an article of clothing will meet this need. This, in turn, necessitates an understanding of the fundamental principles of clothing performance. Toward this aim, we have established that

1. the optimal ambient or microclimate temperature in which humans are able to maintain body temperature without stress is from about 26 to 30 degrees Celsius (79 to 86 degrees Fahrenheit) in air, and
2. the physical laws of convection, conduction, radiation, and evaporation clearly define the routes of heat loss from the body.

The objective of clothing is to achieve the former (1) by manipulation of the latter (2). In general this is achieved by modifying the

▶ volume, characteristics, and number of air-containing compartments within the microclimate;
▶ rate of exchange of microclimate air with environmental air; and
▶ amount of moisture within the clothing.

Clothing for Cold Conditions

Cold environments can be wet or dry, still or windy. For each of these conditions there is a protective antidote. For cold there is insulation, for wind there is windproofing, and for wet there is waterproofing.

Insulation

Clothing insulation works by acting as a thermal barrier between the skin and the environment (figure 3.1). The warm air contained within the clothing, together with radiant heat transfer from the skin, tends to raise the temperature of the inner surface of the clothing fabric. The actual temperature of the fabric will be a balance between the heat gained from

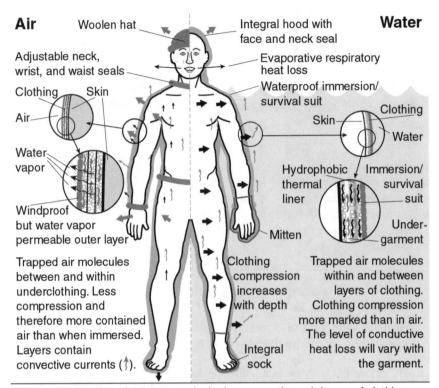

Air   Woolen hat   Integral hood with face and neck seal   **Water**

Adjustable neck, wrist, and waist seals

Evaporative respiratory heat loss

Clothing   Skin

Air

Waterproof immersion/survival suit

Clothing

Skin

Water

Water vapor

Hydrophobic thermal liner

Immersion/survival suit

Windproof but water vapor permeable outer layer

Mitten

Under-garment

Trapped air molecules between and within underclothing. Less compression and therefore more contained air than when immersed. Layers contain convective currents ($\uparrow$).

Clothing compression increases with depth

Trapped air molecules within and between layers of clothing. Clothing compression more marked than in air. The level of conductive heat loss will vary with the garment.

Integral sock

➡ Radiant heat loss progressively decreases through layers of clothing.
➡ Conductive heat loss also less than in naked man.
⤳ Convective heat loss minimized by clothing.
→ Respiratory evaporative loss unchanged. Some still occurs through water vapor permeable clothing.

Figure 3.1   Routes of heat exchange in clothed person in air and water.

the body and the heat lost to the environment from its outer surface. The molecules of warmed air retained adjacent to the skin by the clothing help to protect the boundary layer, maintain skin temperature, and minimize the gradient for heat loss from the deep body tissues to the skin.

As a general rule, because still air is an effective insulator, clothing insulation depends more on clothing thickness and the consequent volume of trapped still air ("dead" air) than it does on the nature of the fibers from which the clothing is made. Thus, clothing does not have to be heavy in weight to insulate, but it must contain trapped dead air to be effective. Clothing insulation is often measured in *clo*.[1] In a still-air environment, clothing insulation provides about 1.6 clo per centimeter (4 clo per inch) of thickness when the air in the clothing is confined to narrow spaces and is therefore "dead."

Although the material from which clothing is made is relatively unimportant to its ability to insulate, its response to compression can make a significant difference. If the material is easily compressed and stays deformed, much of the insulation will be lost with compression as the insulating air is squeezed out of the garment. A strong wind can act as the compressive force. Thus although cotton, wool, and down can provide equal levels of insulation in an uncompressed state, the springy down provides more insulation during and immediately after compression.

As garments increase in insulation they become thicker and bulkier. Bulk sets a practical upper limit for insulation thickness at about three centimeters (one and one-quarter inches), or about five clo. Sleeping bags can provide greater insulation (up to eight clo). Cool environmental air can bypass clothing insulation if it moves in and out of the clothing through the wrist, ankle, and neck openings during activity *(bellows effect)*. Most garments provide adjustable apertures in these parts to prevent or facilitate this effect and consequent heat exchange by forced convection. Occasionally, it may be beneficial to use the bellows effect to assist thermoregulation. Such behavior is especially advantageous when wearing a waterproof outer garment. Forced convection can facilitate the ventilation of undergarments and prevent the buildup of moisture vapor from sweating or insensible perspiration (see page 45).

The thickness of insulation is best provided by layers of clothing rather than one garment. Layered clothing provides a greater obstacle to heat loss and has a weight advantage over a single thicker garment, because lighter materials can produce similar insulation. Layered clothing also provides the option of removing layers to facilitate heat loss during periods of increased activity. Opening clothing apertures can further enhance this heat loss, particularly if the undergarment adjacent to the skin is made from a thick open-weave material to facilitate the circulation of cool air (convection). In cold, windy conditions, where forced convection may be a problem rather than a benefit, an outer windproof layer of clothing will reduce convective heat loss. Behavior can also reduce convective heat loss. A person can reduce body surface area in

---

1. One clo is equivalent to the amount of insulation required to keep a seated subject comfortable in air at a temperature of 21 degrees Celsius (70 degrees Fahrenheit), relative humidity of less than 50 percent, and air movement of 0.1 meters per second (0.2 miles per hour). A 1940s business suit provides about this amount of insulation. 1 clo = $0.18°C/m^2/h/$ kcal = $0.155°C/m^2/W$. The textile industry use the *tog* as the unit of insulation; 1 tog = 0.645 clo.

contact with the air by keeping the legs together and the arms by the sides or by squatting in a fetal position.

Covering the body with clothing is an efficient method of reducing heat loss through radiation, although some radiant heat loss will occur from the outer surface if it is warmer than surrounding objects. Additional insulation will reduce this source of heat loss by lowering surface temperature. Encircling the body in a reflective material can theoretically reduce radiant heat loss (provided the reflective layer faces the body). Some survival aids have used this approach. Although such devices may be of value in warm environments (e.g., hospital wards), they are of little use in cold environments. Two main reasons account for this:

1. When skin temperature approaches ambient temperature, heat loss through radiation is minimal.

2. Water vapor in the microenvironment from insensible perspiration will quickly condense on the cool inner surface of the reflective blanket and reduce its reflectivity to almost zero.[2]

It is far better to enclose the individual in some appropriate vaporproof and windproof barrier (e.g., a heavy-duty polyethylene bag) to help prevent continued heat loss through convection and evaporation (see chapters 9 and 11).

Vapor Permeability

The term *vapor permeability* describes the ease with which water-vapor molecules can pass through a material. The higher the permeability, the greater the volume of water vapor that can pass through the clothing and leave the microclimate. Thus, the vapor permeability of a clothing assembly can significantly influence undergarment wetting and therefore insulation and comfort.

As we have seen, insensible perspiration continues even in a cold environment. This, plus sweating during exercise, are the usual sources of the moisture content within the microclimate of waterproof clothing. This moisture content is related to relative rather than absolute humidity. At cold temperatures, the water content of air can be low but the relative humidity may be high. For example, at 0 degrees Celsius (32 degrees Fahrenheit) the moisture content of air is low, but the relative humidity is about 80 percent (this is one of the reasons why people feel cold in damp weather). Alternatively, at higher temperatures the water content may be high but the humidity low.

Thus, the air in clothing adjacent to the skin can be warm but have a relative humidity of only 50 percent. As that air and the moisture it con-

---

2. Such survival aids were of little use to *Estonia* survivors who tried to use them.

tains permeates through the clothing toward the cooler outer layers, the temperature will fall and the relative humidity will rise until eventually water condenses in these layers. In subzero environments, moisture from sweat can freeze in the outer layers of clothing, producing a layer of ice within it. This ice, by replacing air, reduces insulation and increases conductive heat flow through the clothing. In less severe environments underclothing absorbs the condensed moisture, becoming damp and replacing air. Because water has 24 times the thermal conductivity of air, the displacement of air from clothing by water will quickly and significantly reduce the effectiveness of clothing insulation.

Thus, dampness facilitates heat transfer across the layers of wet clothing through conduction and subsequent heat transfer from the surface of the clothing by evaporation, convection, and radiation.

The wisdom of the old adage "If you want to stay warm in the mountains, stay slightly cold" should now be apparent. By staying slightly cold, a person will minimize sweating and preserve clothing insulation. To prevent the condensation of water vapor within clothing, many manufacturers have replaced impermeable materials with those that incorporate vapor-permeable ("breathable") membranes. When new and clean, these membranes allow water vapor to pass through the clothing while preventing the entrance of water.[3] At least some materials appear to retain this capability when soiled, although salt water, oils, and other substances may degrade other materials.

The ideal garment for cold, dry conditions is windproof but vapor permeable. Even when wearing such a garment, one should carefully manage clothing insulation levels to avoid sweating in a cold environment. If the garment worn is vapor impermeable, the person should ventilate the microclimate periodically to reduce the buildup of water vapor. Although this practice will, temporarily, compromise the defense against convective heat loss, it will be beneficial in the longer term.

Waterproof and Windproof Garments

Water may also enter the clothing from the environment. In wet conditions, an outer waterproof fabric will help preserve the insulative integrity of the undergarments. Manufacturers often combine this feature with windproofing. The value of such protection can be seen in the fact that clothing insulation is reduced by 30 percent in a wind of 14.5 kilometers (9 miles) per hour, by 50 percent by wetting, and by 85 percent by a combination of wind, wetting, and exercise (bellows effect). A dark outer

---

3. Garments permeable by water vapor are usually capable of coping with the small amounts of water vapor produced from insensible perspiration at rest, but few are capable of coping with the evaporation of sweat produced by overheating during work. The efficiency of these membranes declines with age and, in some, with soiling.

Immersion dry suit testing in a laboratory environment.
Photograph courtesy of Frank Golden and Michael Tipton.

covering of a clothing assembly maximizes heat gain from solar radiation. This gain can be significant even in polar regions.

Immersion Clothing

Suits designed for protection against immersion are defined as *wet* or *dry. Wet suits* are normally constructed from closed-cell neoprene three to six millimeters thick. This construction provides the necessary insulation by trapping air within its cellular structure. Wet suits do not have waterproof seals. Instead, they are designed to allow a small amount of water to enter between the suit and the skin. Sizing is therefore important with such suits. A good-fitting wet suit will only allow water to seep between the skin and the suit, where the body will warm it so that it becomes part of the boundary layer and thereafter will not adversely affect insulation. A poorly fitting or damaged wet suit will allow water to flush in and out of the skin-suit interface, constantly disturbing the boundary layer. This action will increase body heat loss (forced convection) and negate the insulation provided by the neoprene. This situation is analogous to air pumping in and out of loose-fitting clothing (bellows effect), but because the fluid is water the effect is even greater.

Dry suits can be uninsulated or insulated. *Uninsulated suits,* as their name implies, have little inherent insulation. They are usually constructed from a trilaminate waterproof material composed of a synthetic membrane bonded to two layers of nylon facing fabric. Often, a breathable, waterproof membrane is bonded to the nylon facing fabrics. These suits usually incorporate waterproof zips and wrist, neck, and face seals. Uninsulated suits are designed to keep the insulation of the clothing worn beneath them dry. The insulating clothing is often normal clothing, which will be adversely affected by water leakage into the suit. As little as half a liter (one pint) of water can produce a 30 percent reduction in the insulation provided by such clothing. To reduce the impact of leakage, one can wear a thermal liner with the suit. Constructed from a water-resistant *(hydrophobic)* insulating material, these tend to cover the torso and upper part of the limbs.

Alternatively, a person can wear an *insulated immersion suit.* These suits are made from material with inherent insulation, such as closed-cell neoprene. Provided water is not flushing in and out of the suit, leakage will have much less effect on an insulated suit than it will on an uninsulated suit worn over everyday clothing. Unlike wet suits, insulated suits are faced with waterproofed nylon fabric and incorporate watertight zips and seals. Some suits provide inherent insulation by inflation of a chamber in the suit with air or carbon dioxide. This approach has the ergonomic and thermal advantage that the user deploys the extra insulation only when needed. In theory, but often not in practice, the use of such insulation, plus breathable membranes and breathable neoprene, will improve thermal comfort of the suits when worn in warm air.

Although important, avoiding or minimizing undergarment wetting is not the sole determinant of retained insulation. With submersion, the pressure of the water surrounding the body *(hydrostatic pressure)* will compress clothing, displace air, and, consequently, reduce insulation. This, along with the differing physical properties of air and water, explains why clothing assemblies have lower clo values in water *(immersed clo)* than in air.

Researchers (Hayward et al. 1978; Hayes and Cohen 1987) have recorded and estimated average cooling rates, immersed clo, and survival times associated with the different types of immersion suit. Results for a thin adult male wearing different clothing assemblies during immersion in calm water at about 12 degrees Celsius (54 degrees Fahrenheit) are:

▸ Ordinary lightweight clothes—cooling rate 2.3 degrees Celsius per hour (4.2 degrees Fahrenheit per hour), immersed clo 0.06, survival time 65 minutes

▷ Uninsulated dry suit over lightweight clothes—cooling rate 1.1 degrees Celsius per hour (2 degrees Fahrenheit per hour), immersed clo 0.33, survival time 4 hours

▷ Full wet suit, 4.8-millimeter closed-cell foam covering extremities and trunk—cooling rate 0.5 degrees Celsius per hour (0.9 degrees Fahrenheit per hour), immersed clo 0.5, survival time 10 hours

▷ Insulated dry suit, 4.8-millimeter closed-cell foam covering extremities and trunk—cooling rate 0.3 degrees Celsius per hour (0.5 degrees Fahrenheit per hour), immersed clo 0.7, survival time 15 hours

The inherent assumption in these figures is that ethically approved tests of immersion suits accurately indicate their performance in a real emergency[4] and that survival time in cold water is related to hypothermia. We will discuss this, as well as the accuracy of the prediction of survival times, in chapter 7. Finally, no matter what clothing one wears and what weather conditions prevail, because the cooling power of water is much greater than that of air, an immersion victim is always better off out of the water than in it, even when it feels colder in the air (Steinman et al. 1987).

Head Protection

Because of the absence of cold constrictor fibers in the blood vessels of the scalp, head protection is important in the cold. For the same reason, a person can use the head to vary overall clothing insulation for thermoregulatory purposes when the removal of other items of clothing may be inconvenient. In the cold, heat loss from the head can account for over half the resting metabolic heat production. Removing head protection will therefore significantly increase heat loss. In extreme cold one should remember that the face is part of the head and that heat loss from the respiratory tract and face can be up to one-third of total heat production. A person should protect these areas, as well as the eyes, but ensure that respiration is unimpeded. To this end, various respiratory heat exchangers have been designed, though without a great deal of success.

Hand and Foot Protection

The hands and feet do not produce much heat. Thus, the temperature of these areas depends primarily on the heat delivered by blood flow. When cold conditions reduce this blood flow by vasoconstriction, the hands

---

4. The validity of the various international standards for immersion suits, life jackets, and life rafts and of the tests, both human and manikin, used to assess them, is outside the scope of this text. For further information on this topic we refer the interested reader to the bibliography at the end of this book.

and feet cool quickly because of their high surface area. The best way to protect the hands and the feet, besides insulating them, is to keep the body warm and thereby maintain their blood (heat) supply. The saying "If you want to keep your hands warm in the cold, wear a hat" is accurate. A hat will reduce heat loss from the head (see the preceding) and help maintain deep body temperature and thus peripheral blood flow.

Gloves and footwear will insulate but will not maintain local blood flow if deep body temperature is falling. Hence, even with gloves and special footwear, if deep body temperature falls, hand and foot temperature will fall, albeit more slowly with extra insulation. The hands are particularly difficult to protect. As the thickness of gloves increases so does the surface area for heat loss; each individual finger represents a cylinder with a high surface area. Fingerless mitts are therefore preferable to gloves. Heated gloves are becoming more widely available. Although these may raise skin temperature locally and thereby improve thermal comfort, their danger lies in the fact that they are heating relatively bloodless tissue, which can be harmful to the tissues.

Conductive heat loss in an upright person is limited to heat transfer through the soles of the feet to the cooler ground, and hence is small in terms of total body heat loss. Nevertheless, good insulating footwear is desirable in cold environments. In normal circumstances, blood flow to the feet of an upright person is poor because of gravitational effects on venous blood flow. In cold climates, blood flow decreases further in an effort to conserve heat, resulting in local tissue temperatures that are usually significantly cooler than elsewhere in the body. The feet are thus prone to cold injury (see chapter 10). Therefore, in such climates one should insulate the feet well and adopt procedures to facilitate an intermittent through-flow of warm blood by exercise and, if possible, periodic elevation and rewarming.

Clothing for Hot Conditions

In hot conditions, clothing should protect the skin from radiant heat while impeding heat loss as little as possible. The "solar toupee," once popular for protecting the brain from the dangerous "skull-penetrating rays of the sun," is no longer thought necessary! However, clothing should protect against sunburn, be light in color or reflective to combat radiant heat gain, and be light in weight (open weave) to facilitate convective cooling and minimize burden. It should be loose fitting to promote air movement next to the skin (bellows effect) and sweat absorbent to wick moisture away and reduce humidity in the microclimate. Although of less benefit for thermal balance than direct evaporation from the skin, the absorption of sweat will improve thermal comfort and facilitate evaporation from the surface of the clothing.

Somewhat in contrast to this advice, populations indigenous to desert environments often wear bulky, voluminous clothing. Such clothing is designed to provide protection against direct and reflected solar radiation while facilitating the bellows effect. The bulky clothing also provides protection against the cold desert nights and the frequent sand storms.

As we have just seen, it is not difficult to find contradictions in the field of thermal physiology. For example, in previous passages we have spoken of immersion suits that will keep a person warm in water but make the person hot and uncomfortable in air, of keeping cool in the cold to preserve deep body temperature, of staying warm to maintain peripheral temperatures. These contradictions are the inevitable consequence of the way the thermoregulatory system works—varying peripheral insulation to help maintain deep body temperature—and the fact that in many survival situations we are placing that system in an environment for which it was not designed.

Chapter Summary and Recommendations

▸ Body temperature regulation by physiological measures alone permits long-term survival only in a tropical environment.

▸ Humans, by use of their intellect and resultant behavior, have learned how to maintain their micro- and macroenvironments in the thermoneutral range. This ability has given them access to the whole of the planet and to space.

▸ The behavioral thermoregulatory responses, driven by thermal comfort, are thus the principal methods that enable humans to survive in the thermally hostile environments outside the tropics.

▸ Knowledge about how best to use technology and behavioral activity can be essential to successful survival. Protection is the cardinal principle of survival, and protection against the thermal threat, cold in particular, must therefore assume a very high priority.

▸ Without such knowledge, your ability to improvise and optimize the use of available materials may seriously compromise your survival.

▸ Equipped with the information provided in this chapter, you should be in a position to select appropriate clothing and use the resources at your disposal to maximize the chances of survival.

C H A P T E R

4

Initial and
Short-Term Immersion

" **I**N 1979 TWO HH-3F HELICOPTERS** ditched in water at temperatures of 13 degrees Celsius (55 degrees Fahrenheit) and 14 degrees
Celsius (57 degrees Fahrenheit). Of a total of nine crewmen, only three
survived. None of those who perished had injuries extensive enough to
have prevented their escape from the inverted, floating craft. The
postcrash investigation revealed that all victims had drowned while
attempting to egress. In each accident, decreased breath hold time,
due to the effects of sudden immersion in cold water was implicated as
the likely precursor to drowning."

Eberwein 1985

By 1985 the initial responses to immersion in cold water and the associated threats were known and had been reported in the scientific literature. But it was unusual for an accident investigation to recognize formally
the part played by these responses in fatalities. Because of other accidents and a large amount of research, the initial responses to immersion
are now more widely regarded as a significant threat to all who are
immersed in cold water. Sadly, however, this fact is still unrecognized in
many fatal accident inquiries and by many engaged in water-based leisure
activities.

Beginning with thermoneutral immersion and progressing to cold, in this chapter we examine the physiological responses evoked when the environment surrounding the body changes from air to water.

IMMERSION IN THERMONEUTRAL WATER

As we saw in chapter 2, although both air and water are fluids they have quite different physical characteristics. In air at sea level, the body is subjected to one atmosphere[1] of pressure and gravity. In water, the effects of gravity lessen significantly, while the pressure rises by one atmosphere for every 10-meter (33-foot) increase in depth *(hydrostatic pressure)*. The effects of external water pressure on the resilient body tissues and gas-containing body cavities account for the physiological changes that occur when an individual moves from being immersed in air to being immersed in water. This effect is greater when the body floats vertically and least when it floats horizontally.

Primarily because of the high density of water, head-out immersion in thermoneutral water (35 degrees Celsius; 95 degrees Fahrenheit) can produce profound physiological alterations in the circulation, heart, lungs (respiration), and the gut. Because of its value as a model for kidney function, the physiological effects of immersion in thermoneutral water have been extensively investigated in recent years and are now well understood (Epstein 1978). When breathing atmospheric air during head-out immersion, the lung of an immersed individual is open to the atmosphere. Consequently, the pressure in the lung is the same as that in air (that is, one atmosphere). The pressure outside the chest wall and remainder of the body, however, is at one atmosphere plus the additional hydrostatic pressure (figure 4.1). This pressure increases by one 10th of an atmosphere per meter in depth. The result is that the lung is at relative negative pressure with respect to the rest of the body. The immersed individual is, in effect, negative-pressure breathing. Although the pressures involved are small in absolute terms, they are well in excess of the pressure of blood in the veins *(venous pressure)*. Consequently, hydrostatic pressure has an important physiological effect. It acts like a jet-fighter pilot's G suit, squeezing the venous blood out of the lower parts of the body back toward the central circulation and heart.

Within six heartbeats of a head-out upright immersion, an increase of 700 milliliters (three-quarters of a quart) in the return of venous blood

---

1. One atmosphere is the pressure exerted by the weight of air in the atmosphere at sea level = 760 mm Hg, 101.3 kPa, 14.7 psi.

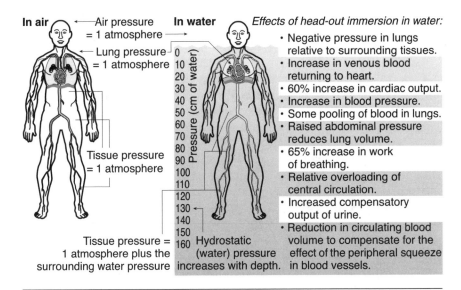

In air   ←— Air pressure   **In water**   *Effects of head-out immersion in water:*
= 1 atmosphere —→
←— Lung pressure   • Negative pressure in lungs
= 1 atmosphere   relative to surrounding tissues.
• Increase in venous blood
returning to heart.
• 60% increase in cardiac output.
• Increase in blood pressure.
• Some pooling of blood in lungs.
Tissue pressure   • Raised abdominal pressure
= 1 atmosphere   reduces lung volume.
• 65% increase in work
of breathing.
• Relative overloading of
central circulation.
• Increased compensatory
output of urine.
• Reduction in circulating blood
Tissue pressure =   Hydrostatic   volume to compensate for the
1 atmosphere plus the   (water) pressure   effect of the peripheral squeeze
surrounding water pressure   increases with depth.   in blood vessels.

Pressure (cm of water): 0 10 20 30 40 50 60 70 80 90 100 110 120 130 140 150 160

Figure 4.1 Diagram illustrating the effects of hydrostatic pressure on the body.

to the heart leads to a 32 to 66 percent increase in the volume of blood being pumped by the heart *(cardiac output)*. The redistribution of fluids results in some pooling of blood in the lungs and consequent encroachment on air space. A reduction in lung volume from hydrostatic squeezing of the abdomen (resulting in upward displacement of the diaphragm) and chest compounds the situation. Overall, the reduction in lung volume results in a 65 percent increase in the work of breathing. Pressure receptors situated in the large blood vessels near the heart detect the increase in the volume of blood in the center of the body and sense it as overloading of the circulation. In an attempt to lower pressure, the body removes some of the circulating fluid by increasing the production of urine *(diuresis)*; it is not unusual for the body to produce 0.35 liters (two-thirds of a pint) of urine within an hour of immersion. This action helps return the blood volume and pressure in the center of the body to normal levels, and the body adjusts to its new aqueous environment.

From a practical viewpoint there is little evidence that these changes cause any respiratory or circulatory embarrassment in a fit individual. They may cause problems, however, for someone with a failing heart. During rescue after protracted immersion in warm water or relatively shorter immersion in cold water when hypothermia may be present, a collapse in blood pressure may occur (see chapter 11).

IMMERSION IN COLD WATER

Water cooler than thermoneutral water temperature for a naked human covers most of the planet. Indeed, most seas outside the Tropics are always colder than 20 degrees Celsius (68 degrees Fahrenheit).

Immersion-related death (of all types, e.g., open water, domestic pools and public baths, etc.) is the third most common cause of accidental death in many countries and worldwide accounts for about 140,000 deaths each year. Since ancient times, perceptive observers have recognized that cold, as distinct from drowning, is a cause of death in those immersed in water. Herodotus (450 B.C.), in his account of the ill-fated seaborne expedition of the Persian battle fleet against Athens, clearly distinguishes drowning from hypothermia. He describes how "those who could not swim perished from that cause, others from cold."

Since those early days and throughout the succeeding centuries, the perceived threat associated with immersion in cold water has changed. Often this change occurred more slowly than it might have because

Old-fashioned flotation device circa 1854.

Photograph courtesy of the Royal National Lifeboat Institution.

people missed vital clues, ignored them at inquiries, or simply forgot the lessons of history. Always, however, the protective equipment provided has reflected what people believed to be the major risk at the time. Thus, in 1757 Col de Galacy, a Frenchman, built a cork life jacket in the belief that the major problem associated with going into water was staying afloat. This belief persisted, and the cork life jacket invented by Captain Ward in 1854 was used up to and during World War II. (The reader particularly interested in the history of the development of the life jacket should refer to Brooks 1995.)

The belief that staying afloat was the major challenge persisted in the face of some convincing evidence that cold also represented a major hazard. For many years the authorities ignored the evidence, such as the sinking of *Titanic,* in which those who entered the calm, cold water perished despite being kept afloat by life jackets.

With time, people began to recognize cold as a major hazard. Many accounts by survivors of maritime tragedies describe how people survived the incident that caused them to enter the water, only to perish in the succeeding minutes or hours before rescue. In the more dramatic incidents, some individuals survived for many hours in very cold water. The majority of such survivors were well endowed with a generous layer of subcutaneous fat. The benefit of this fat lies in the insulation it provides against cold rather than any assistance it might give to staying afloat.

In the 1950s, research by Pugh and Edholm (1955) on open-water long-distance swimmers attempted to discover why such individuals could survive in cold water for long periods, which, according to data obtained during World War II, would incapacitate clothed shipwreck victims. Swimmers who crossed the English Channel frequently spent 12 to more than 20 hours swimming in water of 12 to 15 degrees Celsius (54 to 59 degrees Fahrenheit), although the estimated survival time in those temperatures is about 6 hours. The unusual capability of these swimmers was attributed to their combination of large deposits of subcutaneous fat coupled with high levels of physical fitness.[2] Their fitness enabled them to maintain high levels of heat production for protracted periods, and their subcutaneous fat helped them retain much of the heat generated within their bodies.

Pugh and Edholm's control subjects were nonhabituated and of average fatness. The rapid cooling observed in these subjects supported the conclusion, based on data obtained during World War II, that hypothermia

---

2. Modern distance swimmers tend to be slim and fit athletes who survive largely by being able to cover the same distances as their fatter predecessors in about half the time. In the process they generate much more body heat through their higher work rate and use some of the outer portion of their muscle bulk as insulation.

represented the major hazard to be faced by normal individuals during immersion in cold water (figure 4.2). This finding led to the generally held belief that the important factor in survival in cold water is adequate insulation to reduce the rate of body heat loss and thereby prevent or delay the onset of hypothermia.

The preoccupation with hypothermia and protection from it has persisted up to the present day. This concern is reflected in the output of the media and consequent understanding of the general public; the claims made by immersion-suit manufacturers; the guidelines, specifications, and regulations for immersion-protective equipment; the search and rescue policies of various organizations; and various publications on survival in cold water. The importance of hypothermia to the survival story, particularly the wartime story, was rightly emphasized. But as often happens in a topical and popular area of research, hypothermia was promoted to a position where the less well informed began to believe that it

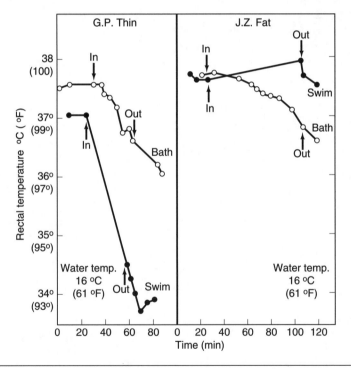

Figure 4.2   Cooling rate of Channel swimmer (JZ) versus control (GP) during a static immersion in a bath (O) and while swimming (●). Note that GP cooled faster with swimming than JZ did and at about the same rate when static, although JZ had much more insulation (fat). In the static immersion GP shivered vigorously, whereas JZ did not shiver appreciably.

Reprinted from Pugh and Edholm 1955.

was the *only* hazard associated with immersion in cold water. Recently a wealth of anecdotal, statistical, and scientific evidence has been compiled that suggests that hypothermia is not the only hazard to be faced. Indeed, it is probably not even the greatest hazard.

Identifying the Major Risk

In medicine, an old adage frequently impressed on young students distracted by the small print of obscure diseases is that "Common diseases occur commonly." The same applies to survival after immersion in cold water. Hypothermia does occur, but the most common cause of death is drowning, and it remains the highest priority for protection. Thus, spending large sums of money on elaborate "survival" suits while ignoring their interaction with flotation equipment, or indeed the effectiveness of such flotation equipment, is foolhardy. Similarly, survival strategies based largely on survival times extrapolated from the cooling rates of fit young volunteer subjects undertaking innocuous laboratory-based tests can be misleading. Such tests, which support the provision of antihypothermia survival suits, may appear reasonable to the casual observer; just as rescue strategies based on expected times to incapacitation from hypothermia (chapters 6 and 7) may also seem reasonable. But neither is of much value if waves are repeatedly breaking over the face of the thermally protected, but drowning, survivor.

Approximately 55 percent of the annual open-water immersion deaths in the United Kingdom occur within about 3 meters (10 feet) of a safe refuge, and about 42 percent occur within 2 meters (6 feet). Two-thirds of those who die are regarded as good swimmers (Home Office Report 1977). Such statistics do not suggest the cause of death to be the protracted period of cooling required for hypothermia. Rather, they are indicative of some incapacitating response that is rapid in onset and prevents individuals from swimming 3 meters to save their lives.

Hypothermia must be kept in perspective. Golden and Hervey did this in 1981 when they described four stages of immersion associated with specific risks:

1. Initial responses (0–3 minutes)
2. Short-term responses (3–30 minutes)
3. Long-term responses (more than 30 minutes)
4. Postimmersion responses (during and after rescue)

The hazards at each of the first three stages are caused by cooling of different regions of the body, starting with the skin (initial responses) and proceeding through the muscles that lie close to the surface of the body, particularly in the arms and legs (short-term responses). In this context, it is worth noting that 50 percent of the tissues of the body are within 2.5 centimeters (1 inch) of its surface (Burton 1935). Finally, the cooling spreads to the organs in the core of the body (long-term responses). In adults, hypothermia cannot occur until stage 3; it is a physical impossibility to lose heat from the surface of the body during immersion at a rate that will cause hypothermia-related problems within 30 minutes. Thus, newspaper reports such as those following the capsize of *Herald of Free Enterprise* (chapter 1) that said, "Following immersion in the icy waters, death from hypothermia occurred within minutes," are misinformed. Before they become hypothermic, immersion victims must survive two hazardous stages. The fatal consequence of failing to survive those stages is usually drowning.

Of all the anecdotal evidence in support of the four stages of risk outlined by Golden and Hervey, none has offered a better illustration than the tragedy that occurred during a short helicopter flight from the oil platform Cormorant Alpha in the North Sea in 1992 (Jessop 1993). Eleven men lost their lives when their helicopter ditched into the cold sea next to the platform it had just left. The comments from the survivors, gleaned from newspaper interviews, were particularly illuminating and give strong support to the stages listed earlier:

▸ Initial response—"The cause of death of those who undid their seatbelts but failed to escape from the helicopter was drowning."

▸ Short-term response—"He was unable to get aboard the life raft."

▸ Long-term response—"He drifted away from the life raft."

▸ Postimmersion responses—"His condition deteriorated significantly during recovery and he died" and "I must have been so relieved I lost consciousness (during recovery)."

Despite this evidence, preoccupation with hypothermia focused the fatal-accident inquiry on those who died at the surface of the sea (long-term responses). The report considered in detail the protection provided against hypothermia by the immersion suits worn. This is as it should be. In contrast, the inquiry did not address the questions of why some survived the impact but failed to escape, and what could have been done about it. Instead, the report covered the problems of these unfortunate individuals by the simple conclusion that they were "overcome by the sea."

It is impossible to say how many more lessons we would have learned over the years and how much better the prospect for immersion victims would have been if after the all too numerous tragedies, the *cause* of the causes of death had been investigated. Words like *drowning, hypothermia,* and *exposure* on death certificates may record the terminal event, but they tell us little about the responses and situations leading up to such deaths. Unfortunately, therefore, little impetus is given to the exploration of measures that could have been taken to prevent them.

Although many still believe that hypothermia is the major hazard that people face on immersion in cold water, the authors of this book believe that the initial and short-term responses are responsible for the majority of immersion deaths in open, cold water. Thus, the wheel has turned full circle. At the beginning of the 19th century drowning was thought to be the principal cause of death following immersion. After a period from the middle of the 20th century when most people believed that hypothermia was the primary cause, drowning is once again considered the major hazard. The difference between then and now lies not in our appreciation of the terminal event but in our understanding of its underlying cause. Although the simple inability to stay afloat is still a factor in nonswimmers, we now emphasize the initial and short-term responses and their role as precursors to drowning. We turn now to these important physiological responses.

Initial Responses (Cold Shock)

The sudden lowering of skin temperature on immersion in cold water represents one of the most profound stimuli that the body can encounter. This circumstance produces a number of significant physiological responses. Each can have a detrimental effect on the body and adversely affect survival chances. The responses, which commence almost immediately upon immersion, peak during the first 30 seconds and last for two or three minutes. The speed with which they occur and the fact that they are not seen on immersion in warm water suggest that they are initiated by cold receptors in the skin and mediated by the nervous system. The superficial location of these receptors helps explain why the body initiates responses so rapidly and why they occur in people who have generous deposits of subcutaneous fat. The water temperature at which the initial responses occur varies between individuals. The responses are most acute in people who are sensitive to the cold and least in those who are cold habituated. Individuals who are unaccustomed to cold water will show the first signs of these responses even in water at 25 degrees Celsius (77 degrees Fahrenheit).

The initial responses, which predominantly affect circulation and breathing (cardiorespiratory), are collectively known as the *cold-shock* response (Tipton 1989, figure 4.3). This response probably accounts for the majority of near-drowning incidents and drowning deaths following accidental immersion in open water below 15 degrees Celsius (59 degrees Fahrenheit).

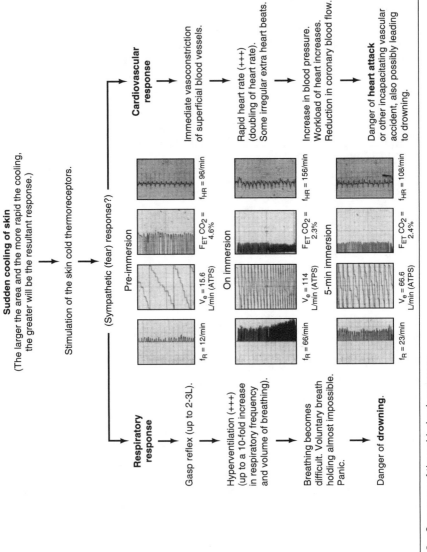

Figure 4.3 Summary of the cold-shock response.

Circulatory Changes

On immersion, immediate vasoconstriction of the skin (peripheral) blood vessels occurs over most of the body. This response increases the resistance to blood flow in the skin and increases flow returning to the heart in the veins because of the hydrostatic squeeze described earlier. In addition, because there is a simultaneous and sudden rise in heart rate, blood pressure rises dramatically. Thus the heart works harder as it tries to pump blood against the raised peripheral resistance.

These responses have an adverse effect on the blood flow to the coronary arteries and can cause problems for those with coronary artery disease. The rise in blood pressure may also pose a problem to those already suffering from high blood pressure *(hypertension)* and can cause a cardiovascular accident (ruptured blood vessel in the brain, or stroke). Hormones *(catecholamines)* secreted into the blood in response to the sudden stress may also produce undesirable abnormal cardiac rhythms in susceptible individuals. Such abnormal rhythms can also occur as a direct result of rapid cooling of the skin and are most likely following breath holding, when a person also immerses the face (such as in snorkeling). These abnormal rhythms are thought to result from competition between the cold-shock response trying to accelerate the heart and the "diving response" (chapter 5) trying to slow it. Face immersion can also lead to a sudden inhibition of cardiac activity, resulting in cardiac arrest *(vagal inhibition)*. These undesirable abnormal cardiac rhythms can cause sudden death, or *hydrocution,* in susceptible individuals.

Incidents of Cold Shock

A 40-year-old man attending a professional crew and skipper course was sailing with two other students when his boat jibed accidentally and he was knocked overboard into the 11-degree Celsius (52-degree Fahrenheit) water. The man, of very large build, was wearing yachting clothing and a 150-newton semibuoyant life jacket. Two instructors in another boat who were accompanying the sailboat were initially able to talk to the student, who reported that he had not been injured. Because of his weight (127 kilograms, or 280 pounds) and the high freeboard of their boat, however, they were unable to recover him from the water. The student, having attempted to assist in his own recovery, appeared to tire. After 4 minutes he suddenly lost consciousness and stopped breathing soon after that. Approximately 20 minutes later a rescue helicopter arrived, rescued, and airlifted the student

(continued)

(continued)

to a hospital. Despite extensive efforts to resuscitate him, in both the water and the helicopter, the student was declared dead in the hospital. After autopsy examination the pathologist, having excluded other possible causes, concluded that the cause of death was hypothermia.

The circumstances of this incident do not suggest hypothermia as the cause of death. The student was large and well clothed and his deterioration occurred too soon after immersion—at a time when his deep body temperature would have been normal, or even rising. The pathologist undoubtedly completed a thorough examination at postmortem, but it is not possible during such an examination to identify electrical abnormalities that have occurred in the heart because of cold shock, the likely cause of death.

In a separate incident, the pilot of a London ship fell from the pilot ladder into the cold water on disembarking. He was quickly recovered, but he died shortly afterward. Evidence of a recent myocardial infarction (heart attack) was found at postmortem and assumed to be the cause of the fall. The alternative scenario that the fall came about for some other reason and that the heart attack occurred because of cold shock was not considered at any stage.

A sudden increase in blood pressure and heart rate should not be a problem for young, fit, healthy people. Their immediate problems will result predominantly from the respiratory component of the cold-shock response.

Respiratory (Breathing) Changes

Sudden cooling of the skin also results in increased respiratory drive on immersion in water below 25 degrees Celsius (77 degrees Fahrenheit). In colder water (cooler than 15 degrees Celsius; 59 degrees Fahrenheit) an initial "gasp" of up to 2 to 3 liters (2 to 3 quarts) in adults precedes uncontrollable rapid overbreathing *(hyperventilation)*. An increase in the rate, rather than the depth, of breathing initially causes this hyperventilation, which can result in a 10-fold increase of the volume of gas entering and leaving the lungs each minute. This amount of overbreathing may cause dizziness and confusion during the first minutes of immersion, although the consequences of hyperventilation in water do not appear to be as severe as those resulting from the same level of hyperventilation in air. The initial gasp increases the lung volume and causes the immersed individual to breathe close to total lung capacity. Even in air, hyperventilating with a full lung is uncomfortable and creates a sensation of breathing difficulty or suffocation. In the first seconds of accidental immersion in cold water, these sensations undoubtedly contribute to feelings of panic.

The maximum breath-hold times of normally clothed individuals fall from an average of over a minute in air to less than 10 seconds on immersion in cold water. Consequently, in choppy or turbulent water where small waves may intermittently submerge the head (and airway), a person has a significant chance of aspirating water during the first few minutes, until he or she can control respiration. Breath holding to facilitate escape from a submerged vehicle, helicopter, or capsized boat may be difficult and result in entrapment and drowning. Even those who manage

Ditched, partially flooded helicopter—one of several scenarios that are particularly hazardous for those with an impaired breath-hold capability.

Photograph courtesy of Frank Golden.

to escape may aspirate some water, a circumstance that can lead to near drowning (chapter 5). The volume required is small, $\frac{1}{4}$ to $\frac{1}{2}$ liter ($\frac{1}{2}$ to 1 pint) for an average individual, particularly when compared with breathing volumes of over 150 liters (5.3 cubic feet) that have been recorded in the first minute of immersion in cold water.

Of all the responses associated with immersion in cold water, we now regard the reduction in breath-hold time on initial immersion as the most dangerous for otherwise fit and healthy individuals.

The Danger of Breath-Hold Time Reduction

In 1973, following the sudden capsize of a boat, eight of the crew were trapped in an air pocket beneath the upturned hull. Although it only involved a short underwater swim to escape, two of the crew were unable to hold their breath long enough to do so and drowned in the attempt (U.S. Coast Guard).

Protecting Against the Initial Responses

Although the initial responses to immersion in cold water have the potential to be extremely hazardous, measures can be taken to avoid, or reduce, the risk they represent.

Avoidance

The best protection is to take all possible steps to avoid falling into the water in the first place. Properly secured and maintained guardrails are an obvious first step. A person in a high-risk situation should wear a safety harness with appropriate length of line (i.e., one that prevents entry into the water). The harness should be clipped to something that will withstand the snatch loading to which it will be subjected. A solo sailor, or one accompanied by family, should ensure that a means of getting back on board exists, is practicable, and is within the capability of those who may be trying to assist in the reboarding.[3]

People should delay abandoning ship as long as possible. When they must do so, they should try to board the life craft without entering the water.

---

3. If you are a family sailor it is a good policy to practice "man-overboard" recovery procedures, when it is safe to do so. This will ensure that those on board will be able to conduct a competent rescue should you be the victim.

Survival of the Fattest?

As already mentioned, most accounts of survival after protracted cold-water immersion occur in people with a generous layer of insulating subcutaneous fat. Being fat, however, does not guarantee survival in cold water and provides no protection against cold shock. The cold receptors that initiate the cold-shock response are exterior to the layer of subcutaneous fat. In contrast, aerobic physical fitness is beneficial because it has a significant negative relationship with cold shock. That is, aerobically fitter individuals have a smaller cold-shock response. Because most people with an overgenerous amount of subcutaneous fat are frequently less fit than their leaner peers, they are likely to have a more marked cold-shock response. The lesson we should learn is to keep fit.

Protective Equipment

For a lightly clothed individual, the initial responses appear to reach their peak in water between 10 and 15 degrees Celsius (50 and 59 degrees Fahrenheit). Below this the water feels colder, painfully so below 5 degrees Celsius (41 degrees Fahrenheit), but the cold-shock response remains at the level it was on immersion in water at 10 degrees Celsius (50 degrees Fahrenheit). The person who must enter the water, or faces a high likelihood of having to do so, should wear a life jacket, as much clothing as possible for added insulation, and a waterproof overgarment. The rate of change of skin temperature (rate of water leakage, the temporal effect) and the number of cold receptors stimulated (skin surface area cooled, the spatial effect) determine the intensity of the cold-shock response. Where possible, therefore, it is preferable to wear watertight clothing that can protect as much of the skin as possible from coming rapidly into direct contact with the water. Should clothing leak a little, it will still provide some protection from cold shock if it slows the rate of ingress of water. The person must be sure to secure the life jacket properly to prevent it from working loose in the water when bobbing up and down with the waves. The individual wearing an orally inflatable or semi-inflatable life jacket should inflate it fully before entering the water because he or she will not have the necessary control of breathing to inflate it during the hazardous initial seconds in cold water. Similarly, the person wearing a manually operated gas-inflated life jacket should inflate it before entering the water, where the cold will affect sensation and manual dexterity and make location and activation of the inflating mechanism difficult. Likewise, a person should not rely on automatic inflation mechanisms—he or she should always inflate manually. Doing this will ensure that the jacket is working properly before entering the water and allow time to take corrective measures if it is not.

Behavior

If a person is not wearing a watertight overgarment, he or she should enter the water slowly to reduce the speed of skin wetting and thus the rate of stimulation of the skin cold receptors. Unfortunately, this is not always possible because gravity rather than personal preference frequently dictates the speed of water entry in man-overboard situations! On entry, the person who experiences cold shock should hold on to something for support until breathing settles down to a controllable level, for perhaps one or two minutes, before attempting to swim even a short distance. If nothing is available to cling to, he or she should stay as still as possible in the water to reduce the rate of displacement of buoyant air from clothing during the critical early minutes of immersion. The person should remember that the cold-shock response will abate after a couple of minutes.

Acclimation

People who work in occupations that involve a high risk of immersion in cold water and who cannot always wear a specialized protective garment might consider undergoing an acclimation procedure. Acclimation or, more correctly, habituation, can significantly reduce the magnitude of the cold-shock response (figure 4.4 and table 4.1).

With repeated immersions (40 min), the reduction in the breathing response is more marked than that in heart rate, with ventilation on day five being almost half of the value seen on day one. By day seven, ventilation is less than a third of its original value (figure 4.4), and subjects experience little distress on immersion. Subsequent experiments have shown that people can reduce the cold-shock response by 50 percent in as few as five two-minute immersions in cold water. Moreover, most of the habituation remains for up to a year. Brief cold showers have a similar but less marked effect. Such habituation probably explains, at least in part, why some people can swim in very cold water without apparent distress or suffering any ill effects. Perhaps Admiral Jellicoe was ahead of his time: "It may be remembered that, Admiral Jellicoe took the wise precaution, while his flag-ship was at Scapa, of subjecting himself to daily cold water baths" (Critchley 1943).

Short-Term Responses

After the skin, the next tissues to be affected by falling temperature are the nerves, muscles, and joints in the limbs. They cool quickly because the limbs have a high surface area, include little heat-containing mass and, because of vasoconstriction, rapidly lose their major source of heat, blood flow.

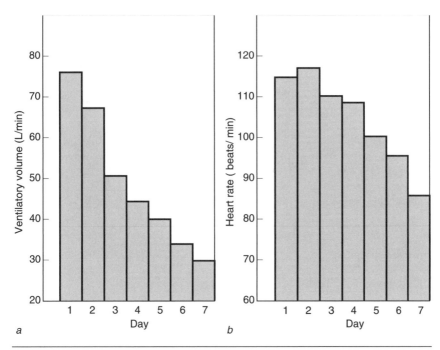

Figure 4.4   Changes in (*a*) ventilatory volume and (*b*) heart rate in seven volunteer subjects during the first minute of daily 40-minute, near-naked (swimming shorts) immersions in water at 15 degrees Celsius (59 degrees Fahrenheit).

Adapted from Golden and Tipton 1988.

In these areas, cold quickly has a debilitating effect on the conduction of nerve impulses, chemical reactions, and muscle mechanics. Skin receptors sensitive to temperature, pressure, and touch lose their sensitivity at a local temperature of 5 degrees Celsius (41 degrees Fahrenheit). The body therefore feels "numb" with cold. The speed with which nervous impulses are conducted in the arm decreases by 15 meters (16 yards) per second for every fall in local temperature of 10 degrees Celsius (18 degrees Fahrenheit). These impulses can be blocked in as little as 1 to 15 minutes at a local temperature of 5 to 15 degrees Celsius (41 to 59 degrees Fahrenheit). The ability of muscle to contract decreases when its temperature falls to about 27 degrees Celsius (81 degrees Fahrenheit); this can happen in the forearm within 20 minutes in water at 12 degrees Celsius (54 degrees Fahrenheit) or within 40 minutes in water at 20 degrees Celsius (68 degrees Fahrenheit). Maximum muscle power output decreases by 3 percent for each fall in muscle temperature of 1 degree Celsius (1.8 degrees Fahrenheit).

Table 4.1   Changes in Initial and Long-Term Responses to Cold-Water Immersion After Eight Daily 40-Minute, Near-Naked (Swimming Shorts) Immersions in Water at 15° C (59° F)

Response	Alteration
First minute of immersion	
Heart rate	↓
Ventilatory volume	↓↓
Respiratory frequency	↓↓↓
Thermal comfort	↑↑↑
Prolonged immersion	
Time to onset of shivering	↑↑↑
Shivering intensity	↓↓
Thermal comfort	↑↑↑

From Golden and Tipton 1987.

Thus, cooling of the peripheral nerves and muscle can quickly cause a level of dysfunction equivalent to peripheral paralysis. This has significant implications for the victim of cold-water immersion. Very soon after immersion, cold can impair the ability to undertake certain activities, some of them critical to survival. In the section that follows we examine two specific examples of this—activities requiring the hands and swimming.

Activities Requiring the Hands

Many activities critical to survival require use of the hands. With time, in some cases just minutes, the ability to use the hands is impaired as they and the muscles in the forearms that help control them cool. Manual dexterity, handgrip strength, and speed of movement can all drop by as much as 60 to 80 percent following immersion. The grip strength of subjects wearing heavy normal clothing can fall by 20 percent after just five minutes of immersion in water at 5 degrees Celsius (41 degrees Fahrenheit). This loss of ability can have serious consequences for activities such as manipulating the oral-inflation valve of an inflatable life jacket, activating a manually inflating life jacket, adjusting a life jacket, tightening straps, deploying a splash guard, tying a buddy line, locating the whistle and other survival aids attached to a life jacket, holding on to flotation aids, boarding a life raft, opening and closing canopy aperture ties, opening the survival contents pack in the life raft, firing a flare, and climbing ladders.

Loss of Manual Dexterity

A pilot attempting to board a ship at night in a 4.5- to 6-meter (15- to 20-foot) swell lost his hold on the ladder and fell between the pilot cutter and the ship. Although he was wearing a personal locator beacon, it did not activate automatically because it was on top of his inflated life jacket and not immersed. The coxswain of the pilot boat, however, was able to maintain visual contact with the life jacket light and an additional strobe light used by the pilot in the water. Unfortunately, because of a design deficiency in the hood of his immersion suit that restricted the pilot's visibility when he deployed it, he was in the habit of wearing his immersion suit with the hood down. Consequently, once immersed the pilot was unable to free the hood from beneath the collar of the life jacket and cold water entered down his neck beneath his "protective" suit. The cold quickly incapacitated him, and he could do little to assist in his recovery (he was even too cold to catch a rope thrown to him). With considerable difficulty, the crew of the cutter were able to recover the cold pilot from the sea.

The impairment in performance caused by such cooling can be spectacular, resulting in adult men being unable to pull themselves out of the water and up a small bank after 10 minutes in iced water. In one experiment, a swimsuited special-forces marine was asked to tear open a soft polyethylene container, remove the dummy flare inside the container, and fire the flare. He was able to do this in seven seconds in air. After 75 minutes in water at 12 degrees Celsius (54 degrees Fahrenheit), his grip strength declined to less than half and he required more than 2 minutes just to tear open the container. With numb fingers he could not manage the twisting action required to fire the flare, despite being in comparatively ideal conditions—seated in calm water and immersed only to chest level. The cause of his difficulty was not his deep body temperature, which was only 2 degrees Celsius (less than 4 degrees Fahrenheit) below normal; the problem was that the muscles and nerves of his hands and forearms had cooled well below 27 degrees Celsius (81 degrees

Fahrenheit). Therefore, he had lost strength, dexterity, and feeling in his hands, which had become like claws.

The problems experienced by the marine were also due, in part, to the effect that temperature can have on the physical properties of material, in this case plastic. In warm air the bag containing the flare was pliant and easily torn. When cold and wet, it became stiff, tough, and slippery. This example highlights a common problem with approved survival equipment. Too frequently it is designed in air-conditioned offices and tested in temperate, unrealistic conditions. Perhaps it is not surprising that people cannot use survival equipment in extreme conditions because of alterations in the physical characteristics of the material used and the diminished capabilities of the user.

One of the most graphic examples of dysfunction and exhaustion caused by cooling of the nerves and muscles comes from the tragedy of the sinking of the MV *Estonia*.

MV Estonia

For most people, the physical effort needed to enter a raft was too much, and after several unsuccessful attempts they were exhausted and gave up. Those in rafts worked hard to bail out the water. Regrettably, through unfamiliarity and cold hands, they could not close the canopy apertures correctly, and the waves therefore washed more water into the raft than they were capable of bailing. In those rafts where the survival-aid container was still present, some survivors described that they were unable to open them to get at the bailer because of cold hands. The plastic wrappings of the flares proved equally challenging to cold fingers. One survivor described how, driven by frustration, he eventually tried using his teeth but gave up when he pulled some out (MV *Estonia*, Final Report 1997)!

The tragedy of *Estonia* represents another example of ignoring the lessons of history. Llano (1955), in a study of over 2,500 incidents, reported that many survivors encountered difficulties while attempting to board life rafts from the water because of exhaustion and cold.

Swimming

Having survived the initial responses, those without flotation aids will be required to make swimming movements to remain afloat or swim to a safe refuge. But it is extremely difficult to swim during the first minutes of immersion in cold water. Many people who are classed as "good" swimmers when swimming in warm water appear to be unable to swim distances of as little as 2 to 3 meters (6 to 10 feet) in cold water, even to save their lives (Home Office 1977).

Swim Failure

In December 1998 five young men were returning from a Christmas dance on the mainland of Scotland to the tiny island of Iona in a 4-meter (14-foot) dinghy powered by an outboard motor. The boat was suddenly swamped and quickly sank in water at about 10 degrees Celsius (50 degrees Fahrenheit). The half-mile crossing was one they had made hundreds of times in the past without incident, so not surprisingly none was wearing a life jacket. All were competent swimmers. One man managed to swim the relatively short distance to shore and raise the alarm. The remaining four died. Sadly, the tragedy resulted in the small community of 80 losing all but one of its men aged 19 to 35.

In an experiment, 10 fit, competent, and fully clothed swimmers were able to complete a 10-minute swim in water at 25 degrees Celsius (77 degrees Fahrenheit). Only 3 of the 10 succeeded, however, when they tried to repeat the swim in water at 5 degrees Celsius (41 degrees Fahrenheit); the remaining 7 failed at times varying between 2 and 7 minutes. At the time of failure, respiratory rate and swimming stroke rate were significantly higher than at the corresponding times of the warm swim. Those who failed had an increase in respiratory rate in the first minute of the cold swim that on average was 122 percent greater than that recorded during the warm swim. The corresponding figure for the three who completed the cold swim was 35 percent. This suggests that swim failure during the first minutes of cold immersion is related to raised respiratory frequency, possibly by making it extremely difficult to coordinate swim stroke with breathing. Normally, a swimming individual will breathe once per swim-stroke cycle, and the ratio of swim stroke to breaths therefore equals one. When respiratory frequency soars on immersion, voluntary control of breathing becomes almost impossible. A large mismatch in the ratio of swim strokes to breathing results, increasing the chances of aspirating water and drowning. Another consequence is that swimming becomes inefficient. The person assumes a more upright position, attempting to keep the mouth clear of the water by using rapid, relatively ineffective paddling movements. In turn, this upright position increases sinking force and drag, necessitating even greater effort.[4] A vicious circle results. The person begins to panic and sink.

The sequence of events just described is thought to explain swim failure during the first minutes of immersion. But swim failure also occurs after the initial responses have disappeared but before the onset of hypothermia. In a more recent experiment, 10 volunteers wearing only swimming shorts undertook self-paced swims for 90 minutes in a swimming flume with the water temperature at 25 degrees Celsius (77 degrees Fahrenheit), 18 degrees Celsius (64 degrees Fahrenheit), and 10 degrees Celsius (50 degrees Fahrenheit). All were very competent swimmers. In the 10-degree Celsius (50-degree Fahrenheit) water, one swimmer reached swim failure in 61 minutes. Four, close to swim failure, were withdrawn before 90 minutes with deep body (rectal) temperatures of 35 degrees Celsius (95 degrees Fahrenheit). For the subjects who failed or came close to failure, impending swim failure was characterized by an increase in stroke rate, a decrease in the length of each stroke, and a dramatic decrease in the distance covered per unit of oxygen consumed.

Toward the end of the swim, shivering coexisted with swimming. The shivering impaired movement and caused postural fatigue, characterized by a painful aching sensation in the back and groin. Swim angle increased as the subjects became more upright in the water, putting them into the previously described vicious circle (figure 4.5). These changes were more marked in the coldest water and in those who came closest to failing. It was also noticeable that as time progressed in the coldest swim, the fingers of all the subjects became more splayed as the cold paralyzed the small muscles of the hand. Swimming style, even in the most proficient swimmers, deteriorated markedly. When close to failure the subjects felt exhausted.

The changes just outlined characterize swimming failure in humans. Although the residual effect of initial hyperventilation may in part cause these changes, they are much more likely to be the result of creeping sensory and muscle paralysis caused by skin and muscle cooling. With cold-induced anesthesia, the loss of feedback from receptors in the limbs, which tell the brain where the limbs are in space, contributes to the deterioration in swimming style. With regard to the fatigue experienced by the swimmers in the cold, muscle blood flow falls with muscle temperature. This means that for the same level of work the blood delivers less oxygen to the working muscle. Consequently, *anaerobic* metabolism

---

4. The specific gravity of water is 1.000, that of the floating body is 1.038, and that of the head alone is 1.110. When the head is immersed, the difference in specific gravity between the water and head is practically insignificant. Lifting the head out of the water, however, changes the swim angle and increases the sinking force of the body. The more of the body that is out of the water, the harder the person must swim to support it.

Figure 4.5 Signs of impending cold-water swim failure. (*a*) Normal swimmer: relatively horizontal in the water; full swim stroke with synchronized symmetrical arm and leg action, both synchronized with breathing; fingers together. (*b*) Onset of swim failure: more vertical in water (about 70 degrees from horizontal); noticeable increase in both stroke and breathing rates; shorter and less efficient leg and arm action, both becoming asynchronous; fingers beginning to splay apart and more flexed; head held higher in the water as effort is made to keep mouth clear of surface; becoming distressed. (*c*) Failure imminent: almost vertical in the water; struggling to keep mouth clear of the surface (neck slightly hyperextended, mouth open gasping for air); ineffective, asynchronous arm and leg action; hands at or near the surface, often splashing; signs of panic; usually unable to lift arm clear of the surface to wave or control breathing in order to shout for help.

(without oxygen) supports a greater percentage of the work undertaken by the muscle. A product of this type of metabolism is lactic acid, and its presence in the muscle causes fatigue and cramping. The coexistence of shivering with exercise in the cold impairs performance and increases the energy consumed for a given amount of external work. Additionally, as with grip strength, the impairment of muscle function by cooling decreases the maximum amount of whole-body work that can be undertaken.

Changes Characterizing Swim Failure

An understanding of the changes that characterize impending swim failure is not merely of academic interest. In August 1999 the following story appeared in some of the national daily newspapers in the United Kingdom:

A 41-year-old Mexican woman died after getting into difficulties while attempting to swim the English Channel. She set off on the 34-kilometer (21-mile) crossing with four others, in a sea temperature of 17 degrees Celsius (63 degrees Fahrenheit), each with an escort boat in attendance. After 11 kilometers (7 miles) and about six hours in the water, she got into difficulties and was removed from the water but sadly was unable to be resuscitated.

Although the precise circumstances of her final moments before rescue were not reported, had she exhibited the signs of imminent failure seen in the laboratory experiments described earlier (figure 4.5), and had those in the accompanying boat recognized them, it is possible they could have withdrawn her from the water before she succumbed.

Swim failure occurs much more commonly in clothed people. The following account of a young man who fell overboard from a sail training ship off Liverpool in 1994 illustrates the speed of onset of incapacitation:

The crew were preparing to shorten sail prior to entering harbor when the fully clothed victim, who was not wearing a life jacket or safety harness, fell overboard. Four marker buoys were thrown into the water and the schooner went about immediately. The captain, who kept his eye on the victim, said, "He was within 6 meters (20 feet) of the ship and was conscious when he surfaced, clearing his hair from his face. But within a few minutes he was in difficulty and next I noticed he was face down in the water."

A popular misconception about drowning is that it is caused by the weight of saturated clothing "dragging people under." This belief has led to the misguided advice to undress in the water, an action that reduces total insulation. But because water does not weigh anything in water, it is not this that drags people under. Rather, a loss of buoyancy occurs when air escapes from within the fabric of the clothing. On initial immersion, air contained within clothing provides helpful buoyancy. After a time, which varies from seconds to minutes depending on the clothing worn and movement, the air escapes, thereby lowering the body in the water and reducing the distance from mouth to water. This requires the person to lift the head higher out of the water to breathe, which may cause the person to enter the vicious circle described previously. The restriction to movement caused by the wet clothing is a further encumbrance; it also increases drag if one is swimming. Even small waves on the sea cause a reduction in the mouth-to-water distance to become critical. Subsequent aspiration of water into the lungs will further reduce the buoyancy of the body.

Rescuers describe how the sound of the rescue boat sometimes prompts people in the water to wave. In so doing, they disturb the trapped air from under their waterproof oversuit and suddenly sink (fishers wearing "oilskins" are particularly prone to this problem). Although it is difficult to do, people who find themselves in similar situations should remain motionless and allow the rescuers to do all the work.

CONCLUSIONS

All the responses described in this chapter can occur before the onset of hypothermia. Sudden incapacitation leading to swim failure is one of the major causes of immersion death. The actual cause of death may be drowning, but that may be the end result of a series of physiological responses that prevent breath holding for even a short period or take away the ability to maintain the airway clear of the surface of the water. These physiological responses may also result in a pathological occurrence, such as a heart attack or stroke, which in turn can cause incapacitation leading to drowning. The eyewitness reports of the sudden cessation of movement in some drowning victims are more suggestive of a sudden circulatory malfunction. The fact that aspiration may occur subsequent to a cardiovascular incident may mislead some into concluding that drowning was the only problem requiring treatment or was the cause of death.

Given the preceding, perhaps the statistic quoted earlier in this chapter is less surprising: Approximately 55 percent of the annual open-water immersion deaths in the United Kingdom occur within 3 meters (10 feet) of a safe refuge.

Chapter Summary and Recommendations

▶ The danger associated with immersion in cold water (less than 15 degrees Celsius; 50 degrees Fahrenheit) has long been recognized but often incorrectly attributed solely to hypothermia.

▶ The preoccupation with hypothermia is reflected in many areas, not least in an industry focused on protecting the survivor against hypothermia while overlooking the nature of other threats.

▶ Some people, even good swimmers, become incapacitated and may even die shortly after falling into water. Cold, choppy water significantly increases this risk.

▶ The time scale of such events precludes a fall in deep body temperature as a possible mechanism. These deaths are more likely to be due to cardiovascular problems or drowning, resulting from the cold-shock response or short-term impairment of nerve and muscle function.

Protecting Against Initial Responses (Cold Shock)

▶ Avoid falling into the water in the first place—use a safety harness in high-risk situations. Ensure that others on board are capable of performing a speedy and effective man-overboard recovery.

▶ Keep fit; consider acclimation.

▶ When abandoning, try to avoid entering the water.

▶ If you must enter the water, or are at a high risk of doing so, wear a life jacket and as much clothing as possible. Enter the water slowly.

▶ Try to ensure that the outer layer of clothing is watertight and can protect as much of the skin as possible from coming rapidly into direct contact with the water.

▶ If you are wearing an orally inflatable or semi-inflatable life jacket, inflate it fully before entering the water.

▶ Likewise, if you are wearing a manually operated gas-inflated life jacket, inflate it before water entry, but *do not* inflate if an underwater escape might be required.

▶ Do not wait for an automatically inflated life jacket to operate. Always operate it manually—*not orally!*—preferably before entering the water.

▶ If you experience cold shock on water entry, hold on to something for support until your breathing settles down to a controllable level (one to two minutes) before attempting to swim even a short distance.

If you have nothing to hold on to, stay as still as possible in the water to reduce the rate of displacement of buoyant air from your clothing during the critical early minutes of immersion.

Protecting Against Short-Term Responses

Most of the advice given for protecting against the initial responses also applies here.

a. Be aware that swimming ability may be significantly impaired in cold open water (swimming can also increase the rate of fall of deep body temperature). Do not estimate your swimming capability on what you can achieve in a swimsuit in warm water.

b. Appreciate that you will quickly lose the ability to use your hands. You should therefore complete any essential survival actions that require manual dexterity and strength soon after immersion.

CHAPTER

5

Drowning and Near Drowning

"IT WAS INTENDED TO BE a pleasant, if challenging, Saturday afternoon's sailing in their new Laser dinghy. Now as a recently married couple, they spent their weeks working in a busy office in London and looked forward to their weekends sailing in the relatively protected waters of the Solent, on the south coast of England. They were accomplished sailors who, as childhood sweethearts, had sailed together in all weathers. This Saturday, however, was to be different.

The force 5 breeze freshened to a force 6–7, as the tide turned, and caught them about a half mile off shore in relatively shallow water. In the steep breaking waves, the dinghy capsized. For some inexplicable reason they could not right it. The water temperature was 12 degrees Celsius (54 degrees Fahrenheit) and they were wearing wet suits and life jackets. Yet somehow they became incapacitated. Someone on shore called the rescue services on seeing their plight. After approximately a half hour a helicopter arrived and rescued them. On arrival in hospital they were still alive although barely conscious. Sadly they were both to die from drowning within the next 24 hours."

It should now be clear to the reader that the great majority of those "lost at sea" die through drowning. *Drowning*[1] has been defined as death through suffocation by submersion, especially in water. It is the third

---

1. The derivation of the English word *drowning* is from the old Northern English word *drune*, which was related to old Viking *drukkna*, "be drowned," from the German base of *drink*.

most common cause of accidental death worldwide, the second for children. In the United States, 40 percent of drowning victims are younger than four years old, and the great majority of these drownings occur in backyard swimming pools (Orlowski 1988). Drowning is caused by an inability to maintain the airway clear of the water long enough to breathe normally. Thus, total submersion is not necessary for drowning to occur; a person may aspirate sufficient water during intermittent submersion of the face, such as that caused by wave splash (see chapters 6 and 7).

Contrary to popular belief, the lungs do not have to be full of water to cause death from drowning. It is generally accepted that the lethal dose of seawater for humans is about 22 milliliters per kilogram (0.34 ounces per pound) of body mass (weight) (Modell 1971).

Thus, if the average 70-kilogram (154-pound) person with a lung volume of 6 liters (6.3 quarts) were to inhale *(aspirate)* about $1\frac{1}{2}$ liters (1.6 quarts) of seawater, he or she would be unlikely to survive. In fact, after aspirating as little as $\frac{1}{4}$ to $\frac{1}{2}$ liter ($\frac{1}{2}$ to 1 pint), there is a danger of dying from complications to the normal functioning of the lungs *(near drowning),* unless the correct treatment is initiated. *Near drowning* is defined as survival, at least temporarily, after aspiration of fluid into the lungs.

At the time of rescue, most near-drowning victims are conscious and breathing spontaneously, often coughing or vomiting but not always. Those who are conscious are usually very frightened from their experience; some even feel well enough to reassure their concerned rescuers that they are fine and in no need of hospital attention. Immersion victims who have aspirated water and are coughing, because of the irritating effect of the water on their airways, should *always* be screened to ensure they are not suffering from near drowning. Near drowning is a medical emergency requiring immediate specific treatment to prevent a rapid deterioration into a progressively worsening condition, which can be, and indeed often is, fatal unless treated. These delayed effects are synonymous with the term *secondary drowning* used in some older publications.

Accounts of near-drowning victims reveal a wide variation in their recollections of their near-death experience. Some describe a period of terror while they struggled to hold their breath until they were no longer capable of doing so, and then feeling a tearing, burning sensation in their chests as water entered their airways. In contrast, others describe a feeling of absolute calmness and tranquillity, with panoramic views of their past lives passing before their eyes. Others have observed a terminal period of stimulation of the central nervous system (CNS), resulting in high blood pressure, vomiting, micturition, defecation, seminal emission, and convulsions before eventual CNS depression with deep coma, blood-pressure collapse, slowing of respiration, and death.

These reported differences in sensation may relate to the level of *hypoxia* (shortage of oxygen) present before rescue. Searing pain probably accompanies the entry of water into the airways, but as hypoxia progresses the sense of tranquillity becomes the overriding sensation. Regardless of which sensation dominates, the threat to survival has more to do with the physiological disturbances resulting from the entry of water into the lungs and its effect on their function and that of the heart. To understand the effect on lung function of aspirating water, we must have some basic knowledge of how the lung works.

LUNGS

The lungs consist of two large air sacs into which air passes from the nose and mouth by a series of tubes. These branch, or subdivide, into a vast number of progressively smaller bored tubes *(bronchi),* until eventually they terminate in a myriad of microscopic air sacs called *alveoli* (figure 5.1). The structure of these airways is similar to a tree; in fact it is

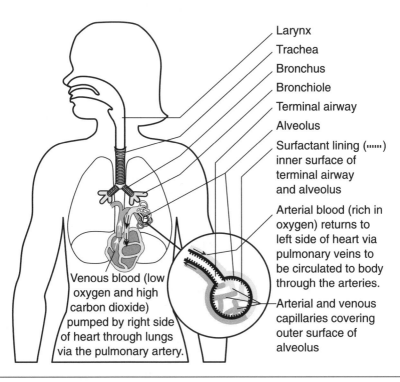

Larynx
Trachea
Bronchus
Bronchiole
Terminal airway
Alveolus
Surfactant lining (••••••)
inner surface of
terminal airway
and alveolus

Arterial blood (rich in
oxygen) returns to
left side of heart via
pulmonary veins to
be circulated to body
through the arteries.

Arterial and venous
capillaries covering
outer surface of
alveolus

Venous blood (low
oxygen and high
carbon dioxide)
pumped by right side
of heart through lungs
via the pulmonary artery.

Figure 5.1   Diagram of lungs, blood circulation through heart, and airway.

often referred to as the *bronchial tree,* with the trunk corresponding to the main airway connecting the mouth and nose to the chest (the *trachea*), and the multitude of branches and subbranches to the different-sized bronchi. The leaves would represent the terminal air sacs, the alveoli. Adults have about 300 million alveoli with a collective surface area of around 60 to 80 square meters (72 to 96 square yards), the area of 20 king-sized beds or half a tennis court. This is the area available for gas exchange with the blood. The outer surface of each alveolus is covered with a film of blood, fed by millions of microscopic blood vessels *(pulmonary capillaries)* and separated from the gas in the alveoli by a very fine membrane with the thickness of one flattened cell. Oxygen and carbon dioxide pass across this membrane (alveolar membrane) at a rate determined by the gas-pressure differentials between the blood and alveolar gas. On inspiration, the air entering the alveolus contains more oxygen and less carbon dioxide than the blood. Because of the pressure differentials, oxygen diffuses across the alveolar membrane into the blood and carbon dioxide diffuses the opposite way. The carbon dioxide (an end product of metabolism) is then exhaled on expiration.

The poorly oxygenated but carbon-dioxide-laden venous blood returning to the right side of the heart from the body is pumped under low pressure through the pulmonary arteries into the lungs, where it picks up oxygen and off-loads carbon dioxide. From the lungs, the freshly oxygenated blood returns to the bigger, stronger left side of the heart. From there it is pumped through the systemic arteries under high pressure around the body, where it provides oxygen to the active tissues before returning once more in the veins to the heart to repeat the cycle.

On inspiration, the chest muscles contract, pulling the ribs up and out, and the diaphragm contracts, pushing the abdominal contents farther down into the abdomen. The result is the creation of negative pressure inside the chest into which air flows freely provided the airway is unobstructed. On expiration, the chest muscles and diaphragm relax, and the elastic recoil of the lungs passively expels air from the chest.

The small terminal airways contain microscopic glands that secrete an oily type of fluid called *surfactant.* Without it the lungs would collapse on expiration and, like a bubble-gum bubble, be impossible to reinflate. The blood pumping through the lungs would then be unable to come into contact with the air. Death would soon follow from lack of oxygen.

Following aspiration of water, the nature and content of surfactant changes, depending on the volume of water aspirated and its type (salt or fresh). Aspiration of a relatively small volume of either fresh or salt water washes some of the surfactant out of the lungs, causing small, localized areas of collapse. *Fresh water,* being less salty than blood, is rapidly absorbed across the alveolar membrane by *osmosis* and taken away in the blood stream. The residual damage to the surfactant leaves some patchy areas of temporary alveolar collapse in an otherwise relatively "dry" lung. This collapse reduces the area of the lung available for gas exchange, which immediately alters the oxygen content of the blood. *Seawater,* on the other hand, because of its increased salt concentration relative to blood, attracts fluid from the blood stream into the alveoli by osmosis, thereby increasing the overall volume of fluid retained in the lung. This results in a "wet" lung, which also reduces the area of the lung available for gas transfer.

In both types of water aspiration, although more so with seawater, the irritation to the lung tissue that has been in contact with water produces a localized *inflammatory response.* This results in an outpouring of fluid from the blood, across the alveolar membrane and into the alveolus and small terminal airways (see also the section on *pulmonary edema* on page 86). The fluid is similar to that seen in a skin blister following a minor burn. The result of aspirating a given volume of fresh water or seawater will be a fall in the oxygen content of the blood, with a significantly larger fall caused by the seawater. As a result, seawater is more damaging and more lethal to aspirate. Indeed, experiments have shown seawater to be about twice as lethal as fresh water (Modell 1971).

When fresh water is added to blood, the blood cells absorb some of the water by osmosis. As a result they swell and rupture, releasing the potassium they contain into the blood stream. If this happens on a large enough scale, the released potassium will cause cardiac arrest. For many years it was assumed that the cause of death in freshwater drowning was cardiac arrest from the increased level of free potassium. For this to occur, however, an approximately equal volume of water must mix with the blood. By the time such a volume of water gains access to the blood stream, cardiac arrest from hypoxia and other reasons (see later sections) will have long since occurred. In practice, both salt water and fresh water cause death through asphyxia.

PRELIMINARY STAGES OF DROWNING

On submersion, a conscious person will usually attempt to breath hold for as long as possible. A shortage of oxygen or a buildup of carbon dioxide does not determine voluntary maximum breath-hold time in air. Instead, the primary influence is a buildup of nerve impulses from the static respiratory muscles to the respiratory control center in the brain. These impulses commence on the cessation of inspiration and steadily increase in frequency as the duration of the breath hold continues until finally the desire to breathe becomes overwhelming. Evidence for this comes from the fact that at the end of a maximum breath hold, the urge to breathe can be temporarily attenuated and breath-hold time extended a little by swallowing or by other actions that move the respiratory muscles. For example, exhaling and rebreathing from an otherwise empty sealed bag allows a person to approximately double the time spent without fresh air. However, if he or she continues to rebreathe from the sealed bag, the levels of oxygen and carbon dioxide will eventually drive respiration. In cold water the sudden stimulation of the skin thermoreceptors produces a drive to breathe independent of the response just described (see chapter 4).

The attempt to extend breath-hold time by swallowing is the probable explanation for the large volumes of water found in the stomachs of some, but not all, drowning victims. Those who manage to breath hold are more likely to swallow, which could explain why water-filled stomachs occur in some but not every case of drowning. When voluntary breath holding is no longer possible, aspiration of water will occur with inspiration and result in either reflex spasm of the throat *(laryngospasm)* or, more frequently, coughing and gagging. Laryngospasm has been postulated as the likely cause of so-called *dry drowning* (see table 5.1), in which no water can be found in the lungs at autopsy. This circumstance has been reported to be the case in 10 to 20 percent of drownings (Moritz 1944). Some suggest that the *diving reflex* may explain dry drowning (see page 88), but a more likely explanation is sudden cardiac arrest or, if some aspiration does occur before arrest, the absorption of that water from the lungs into the blood. In his survey of 3,000 drowning autopsy reports, Fuller (1963) found no cases suggesting dry drowning. Microscopic damage to lung tissue was found in all cases where such examinations had been conducted, indicating that some aspiration had occurred.

Near Drowning

Should rescue occur before the victim aspirates large volumes of water, he or she is likely to be conscious but frightened, agitated, or confused. The victim may also be coughing and vomiting, have a rapid pulse rate, and sense having difficulty in "getting enough air to breathe" despite adequate breathing movements of the chest. Depending on the volume of water aspirated, the lungs will be stiffer and the work of breathing increased. Some victims may exhibit asthma-like symptoms. The lips and fingernail beds may have a blue tinge *(cyanosis)* indicating a serious shortage of oxygen in the blood. The absence of cyanosis, however, should not be interpreted as evidence that the blood is adequately saturated with oxygen. As stated earlier, victims who have aspirated only a small volume of water may not exhibit overt signs of oxygen shortage at an early stage but can still have patchy areas of lung collapse and a significant fall in blood oxygenation. Some may go on to develop more serious, even life-threatening, complications.

Table 5.1   Terms in Common Usage
Associated With Immersion Death and Drowning

Term	Definition or explanation
Near drowning	Survival, at least temporarily, after aspiration of fluid into the lungs, resulting in impairment of gas transfer, hypoxia, and possibly death.
Drowning	Death through suffocation by submersion, especially in water.
Secondary drowning	The delayed onset of the symptoms of near drowning.
Dry drowning	Death after submersion, when no water is found in the lungs at autopsy.
Vagal inhibition	Extreme slowing, or even stoppage, of the heart from stimulation of the vagal nerve by immersion of the face in water.
Diving response	The triple response characterized by reflex breath holding *(apnea)*, peripheral vasoconstriction, and slowing of the heart *(bradycardia)*, characteristically seen in diving animals (e.g., seals).
Hydrocution	The term sometimes given to the sudden death, particularly in fit young people, after immersion, when no obvious cause can be detected at autopsy.
Cold shock	The initial response to sudden immersion in cold water, seen in unhabituated or unprotected individuals, characterized by gasping, rapid uncontrollable breathing, and an increase in heart rate and blood pressure. Cold shock may cause cardiac arrest or produce incapacitation leading to water aspiration.

When the person aspirates a small volume of fresh or salt water, chances are it will be quickly absorbed across the alveolar membrane into the blood. With time, the local inflammation will disappear, the surfactant-producing glands will recover, and full lung function will return.

With larger volumes of aspirated seawater, the local irritating effect of the fluid in the lungs will increase the inflammatory reaction. Inflamed alveoli will contain less gas (including oxygen), and the blood flow in their associated capillaries will diminish. The fall in oxygen content in these inflamed alveoli results in diversion of their blood flow to adjacent undamaged alveoli. Unfortunately, this development can raise the pressure in the capillary bed of the undamaged alveoli, cause leakage of fluid into them, and extend the area of lung damage.

When mixed and agitated with inspired air on breathing, the aspirated fluid and that leaking from the inflamed alveoli form a froth that is often colored pink. The victim may inhale this froth, termed *pulmonary edema,* into alveoli that may otherwise have escaped damage, interfering with their gas-exchanging capability. This occurrence can lead to serious deterioration in the general condition of the victim. The appearance of pulmonary edema is indicative of a poor prognosis unless intensive care begins immediately. Breathing becomes progressively more difficult, and blood oxygenation is increasingly impaired. Eventually, if untreated, victims may drown in their own leaking body fluid.

Thus, although victims may have been rescued and be conscious and breathing spontaneously, there is no guarantee that they will not deteriorate in the succeeding minutes or hours, or even die, if they do not receive correct treatment. Sometimes this effect can be sudden and dramatic, leading to unconsciousness and the appearance of a frothy fluid at the mouth and nostrils. All near-drowning victims, therefore, must be assessed, preferably in a hospital, to quantify the extent of their lung damage and to have access to an intensive-care unit.

Those experiencing breathing difficulties at the time of rescue will obviously be correctly identified as being in the high-risk group. Less easy to identify are those who seem to make a quick recovery and outwardly appear fit and healthy shortly after their rescue. The presence of coughing, chest pain, a crackling, wheezy sound heard when one applies the ear to the chest wall, or other evidence of respiratory distress is an indication for hospital screening. Those not showing these signs, and therefore not perceived to be at immediate risk, should be monitored periodically over the following 12 hours for the appearance of these signs and symptoms. The pulse rate should also be monitored and recorded; a rising pulse rate might be indicative of the onset of complications.

Drowning

In drowning victims, the entry of the water into the lungs displaces air and thereby reduces buoyancy. With subsequent breathing, a volume of water washes in and out of the lung with each respiratory movement, reducing the likelihood of successful resuscitation. Some of this water, even in seawater drowning, crosses the alveolar membrane and enters the blood stream, some remains trapped in the deeper recesses of the lungs, and most is flushed out on expiration along with some surfactant. The water in the lungs makes them heavier, less elastic, and more turgid, somewhat analogous to the change that occurs when a springy dry sponge becomes wet. It also affects the dynamics of the blood flowing through the lungs from the right side of the heart to the left.

In water-saturated lungs, the resistance to blood flow increases significantly. A rise in pulmonary artery pressure results, causing a retrograde increase in pressure on the right side of the heart. This pressure transfers backward to the large veins entering the heart, producing venous congestion and consequent reduction in the volume of blood returning to the right side of the heart, affecting its output. With less blood flowing through the lungs into the left side of the heart, left-side cardiac output and systemic blood pressure fall. Initially, the heart rate increases in an attempt to compensate, but then, with the onset of general circulatory failure, the heart rate starts to slow. The time from aspiration of a significant volume of water to circulatory arrest appears to be about 1 or 2 minutes. Thereafter, some respiratory movements may continue sporadically for a further 5 to 10 minutes, although these are very intermittent toward the end of this period.

Consideration of the above can lead to the conclusion that drowning is a little more than straightforward suffocation (see table 5.1). Because of the severe pulmonary hypertension associated with aspiration, cardiac arrest occurs earlier than would be expected from hypoxia alone. Thus, in drowning, cardiac failure appears to contribute to, rather than result from, the hypoxia associated with suffocation.

Duration of Submersion Compatible With Survival

The difficulty of assessing the likelihood of survival based on eyewitness accounts of probable submersion times is self-evident. Reports of a victim being "in the water for 5 to 20 minutes" do not necessarily mean that the person was underwater throughout that time. Likewise, it does not follow that someone face down in the water is necessarily aspirating water (see the section on diving response on page 88). Similarly, victims trapped in submerged boats or vehicles may have a pocket of trapped air to breathe for some, or all, of their submersion time. Other interacting factors such as the condition of the victim before submersion (conscious, unconscious, injured), age, and water temperature may influence the outcome.

Some authorities state that unless a pre-existing medical event led to the drowning episode, individuals in whom effective ventilation and circulation are restored within three minutes of the start of submersion will have an excellent chance of normal survival. If the submersion time is three to five minutes, survival may still be likely. But the longer the interval from the onset of submersion to the re-establishment of effective ventilation and circulation, the more likely it is that permanent brain damage will occur. With submersion times of five minutes or more, normal

Prolonged survival underwater is more frequently encountered in small children.
Photograph courtesy of Frank Golden and Michael Tipton.

recovery is uncommon. Note that these figures are based on Modell's experience of drowning victims in the relatively warm waters of Florida.

The increasing number of successful resuscitations of victims, in particular small children, rescued from very cold water suggests that some factor or factors associated with lowered temperatures may beneficially influence outcome. Several authors have postulated that the *diving response,* either alone or in combination with hypothermia, may explain these remarkable survival incidents (Gooden 1992).

The *diving reflex,* or more correctly the *mammalian diving response,* is characterized by reflex breath holding *(apnea),* selective vasoconstriction of the peripheral blood vessels, and a marked slowing of the heart *(bradycardia).* Stimulation of the nerves around the eyes on immersion of the face initiates the response, which is enhanced by fear and cold. The diving reflex plays a powerful role in oxygen conservation in diving animals, enabling some to remain submerged without breathing for periods of 30 minutes or more. Although the diving response is stronger in children than in adults, it is still much weaker than in animals.

On submersion in cold water the stimuli for both the diving response and the contrasting cold-shock response are present. Some believe that survival depends on which of these competing responses prevails. In most naked or normally clothed adults, cold shock predominates (Tipton 1989), but about 15 percent of people show some evidence of the diving response. This percentage can increase in situations where the cold-shock response is reduced. Given that the diving response is relatively constant, declining only with age, it is the strength of the cold-shock response in a given individual that is the primary determinant of which response occurs on submersion. Several factors alter the magnitude of the cold-shock response. For example, the number of people demonstrating a diving response rather than a cold-shock response should increase with increasing water temperature as the magnitude of the cold-shock response declines. Cold-habituated subjects or those wearing specialized protective clothing (for example, dry suits, which prevent rapid cooling of most of the skin while leaving the face exposed to the cold) should also have a greater chance of evoking a diving response.

The simultaneous activation of the diving and cold-shock responses can in itself have consequences for survival. In an experiment in which subjects were submerged in water at 5 degrees Celsius (41 degrees Fahrenheit) or 10 degrees Celsius (50 degrees Fahrenheit) wearing dry suits and breathing through a snorkel, 11 of the 12 subjects, in 31 of 36 submersions, demonstrated a variety of cardiac *arrhythmias* (abnormal rhythms). These occurred just before, and within 10 seconds of, the break of breath holding. Such abnormal cardiac rhythms might explain some of the mysterious sudden deaths *(hydrocution)* that occur in apparently fit young adults when accidentally immersed.

We have seen that individuals immersed in cold water who are without the protection of specialized immersion suits or are not cold habituated will invariably undergo a cold-shock response. Why then do some people, especially small children, occasionally survive protracted submersion in ice-cold water?

Drowning in Very Cold Water—With Subsequent Survival

The first account in the literature of such an incident appeared in 1963, when Kvittingen and Naess described the successful resuscitation of a 5-year-old Norwegian boy following 22 minutes of submersion in ice water. Since then controversy has reigned regarding the mechanisms that enable the extension of the hitherto accepted limit of 5 to 7 minutes for cerebral survival from hypoxia. In the interim, many similar cases have been documented (see Golden et al. 1997 for review). Currently, the longest accidental submersion with an intact neurological outcome is 66 minutes in a $2\frac{1}{2}$-year-old girl in Salt Lake City (reported by Bolte et al. 1988). Regrettably, for every successful case reported many more have unsuccessful outcomes.

Trapped Under Ice

In a more recent account, Gilbert et al. (2000) reported the remarkable survival of a 29-year-old woman who had been trapped under ice for 79 minutes in northern Norway after a skiing accident.

"Initially, she appeared to be able to keep her airway clear of the ice-cold water in which her body was immersed, but after 40 minutes she stopped struggling and appeared to lose consciousness. (Her legs with attached skis were protruding from the ice but her companions were unable to help free her.) A rescue team arrived 39 minutes later to cut the woman free. By then she was extremely cold and apparently dead. Nevertheless CPR was started and continued in the air-ambulance helicopter. When she arrived at the hospital, 2 hours and 50 minutes after falling through the ice, signs of life were absent and her deep body temperature was 14.4 degrees Celsius (57.9 degrees Fahrenheit). By the time she was started on heart bypass her temperature had fallen a further 0.7 degrees Celsius (1.2 degrees Fahrenheit) to 13.7 degrees Celsius (56.7 degrees Fahrenheit) before it began to rise slowly. Eventually her heart and circulation recovered spontaneously as she warmed up. After 9 hours of resuscitation she was transferred to the intensive-care unit, where she remained for 28 days. A year later she still had some residual symptoms but was fit enough to return to her medical studies and skiing (Gilbert et al. 2000)!"

The cases of successful resuscitation after protracted cold-water submersion have several features in common. In most cases the victims are small children; the duration of submersion is less than 15 minutes; early commencement of resuscitative efforts occurs; the water temperature in which they were submerged is below 10 degrees Celsius (50 degrees Fahrenheit), usually at or near freezing; and the victims are usually hypothermic on admission to the hospital, some at surprisingly low deep body temperatures for the reported duration of immersion. A history that incorporates some or all of these criteria, however, does not guarantee a successful outcome.

It would be reasonable to assume that brain survival from hypoxia in these cases must be associated with brain cooling. Brain cell metabolism, and thus oxygen requirement, is reduced with cooling. We know that brain survival time from hypoxia doubles when it is at a temperature of 30 degrees Celsius (86 degrees Fahrenheit), and brain activity appears to cease at temperatures below 22 degrees Celsius (72 degrees Fahrenheit). Therefore, if an individual is to survive a period of submersion without sustaining brain damage, rescue must occur within 5 to 7 minutes at normal brain temperature or within twice that range of time at a brain temperature of 30 degrees Celsius (86 degrees Fahrenheit). For submersions of longer duration, even colder brain temperatures will be required. This begs the question, is it possible for the brain to cool to these levels in the relatively short time before circulatory arrest from drowning occurs?

Children, with their relatively large surface-area-to-mass ratio and small amount of subcutaneous fat, are more susceptible than adults to hypothermia through surface cooling during immersion. Nevertheless, in spite of the well-understood physical capacity of cold water to extract heat from the body, it is extremely unlikely that the rate of fall of deep body temperature by surface cooling alone will be sufficient to cool the brain to the level required to protect it. The cardiac surgery anesthetic literature reports that during surface cooling of anesthetized naked infants with ice packs and ice-cold water, deep body temperature decreases by a maximum of 2.5 degrees Celsius (4.5 degrees Fahrenheit) in the first 10 minutes of the procedure. A further 32 minutes pass before the temperature falls to 24 to 26 degrees Celsius (75 to 79 degrees Fahrenheit) (Mohri et al. 1969). Furthermore, for surface cooling to be effective in quickly extracting heat from the body core, good peripheral circulation must be maintained. This circumstance is unlikely because of cold vasoconstriction. Therefore, it seems improbable that surface cooling alone, even in freezing water, will reduce brain temperature by 7 degrees Celsius (12.6 degrees Fahrenheit) and double brain survival time during the initial 10 or 12 minutes of submersion. We must find an alternative mechanism of protection.

The most likely explanation requires rapid and significant heat exchange in the lungs. Drowning is a dynamic process in which, as long as respiration continues, cold water washes in and out of the lungs with every breath. Experiments have shown that after submersion, respiration continues for about 70 seconds (Conn et al. 1995). Although the actual amount of water retained in the lungs and blood may only be a small proportion of the total amount aspirated, the tidal volume flushing in and out of the lungs will significantly influence heat exchange. To reduce the temperature of the brain, the cooled blood in the lungs must circulate to the brain. This requires the maintenance, at least temporarily, of an intact circulation and effective cardiac output.

Consequently, it is possible that in the early stages of submersion brain temperature will be several degrees colder than rectal temperature. Once effective circulation ceases (about one or two minutes), this differential will exist until body temperature slowly equilibrates. After circulatory arrest, the major route of heat exchange is conduction through the tissues to the surrounding water.

Thus, provided that the water aspirated is very cold and cardiac output lasts long enough for sufficient heat exchange to occur, the brain can cool sufficiently quickly to a temperature low enough to protect against hypoxia. The interplay between these factors determines survivability.

This hypothetical model of the mechanisms of body cooling in submersion victims helps to explain why some children survive protracted submersion in cold water without suffering irreversible brain damage from oxygen starvation. In the case of the adult skier described earlier, it is likely that she maintained her airway clear of the surface of the water until she lost consciousness after 40 minutes. At that time she would have been generally hypothermic, probably with a deep body temperature in the low 30s Celsius (high 80s Fahrenheit). At that temperature her hypoxic survival time would be almost doubled to 10 to 12 minutes. With further cooling, this time would progressively lengthen as general metabolism slowed and the requirement for oxygen by the cells decreased. Eventually, metabolic activity would halt, and the requirement to breathe would cease, although cardiac activity could continue for a little longer and assist in transferring heat from the core of the body to the surface. But it is more likely that on losing consciousness she was no longer able to maintain her airway clear of the water and became submerged. At this point she would aspirate sufficient water to cool the brain rapidly, thereby obtaining a greater degree of protection against hypoxia before she asphyxiated. After developing cardiac arrest, she would remain in a state of hypothermic suspended animation until rescued. Although she suffered severe lung damage from the aspirated water, modern resuscitative techniques and a good deal of skill made her successful resuscitation possible.

Unfortunately, rectal temperature on admission to a hospital is not necessarily a reliable prognostic indicator. A small drop in rectal temperature could reflect short-duration immersion, relatively warm water temperature, or an early cardiac arrest. Alternatively, a large drop could indicate prolonged immersion in cold water with or without intact circulation, a late arrest, or continued surface cooling after rescue and during resuscitative measures. The critical body temperature is that of the brain and how quickly it cools to a level that protects it from hypoxia. Rectal temperature may not provide any indication of this because of a possible temperature differential between body mass and brain.

CONCLUSIONS

From the foregoing it is evident that even relatively small amounts of water in the lungs may alter lung function and result in a decrease in the amount of oxygen in the blood. Even if not immediately fatal, water in the lungs, particularly seawater, will damage the local tissues and may prove fatal some hours later. Duration of immersion is not a good indicator of the volume of water a person is likely to have aspirated. Finally, some of those who are apparently dead at the time of rescue after submersion in ice-cold water may respond to resuscitation, particularly when cardiopulmonary bypass is used.

Chapter Summary and Recommendations

▸ Drowning is the most common cause of death at sea.

▸ Continued aspiration of water after submersion usually results in unconsciousness and death from cardiac failure in less than two minutes.

▸ The lungs do not have to be full of water to cause death from drowning. Aspiration and retention of about $1\frac{1}{2}$ liters (1.6 quarts) of seawater is invariably fatal to an adult person.

▸ Aspiration and retention of as little as $\frac{1}{4}$ to $\frac{1}{2}$ liter ($\frac{1}{2}$ to 1 pint) of seawater can cause significant changes to lung function (near drowning) and possibly lead to death from hypoxia.

▸ A person may aspirate such volumes through "wave splash" from waves breaking against the front of a life jacket and washing over the mouth and nose (chapters 6 and 7). Thus total submersion of the head is not necessary to produce drowning.

▸ Rapid cooling of the brain can accompany drowning after submersion in ice-cold water. This sequence of events can sometimes protect the brain from hypoxia. In such dynamic circumstances, rectal temperature is a poor indicator of brain temperature.

▸ The diving response may help breath holding in warmer waters (around 20 degrees Celsius; 68 degrees Fahrenheit). In cold water, however, the cold-shock response usually predominates.

▸ Those suspected of aspirating water and exhibiting some evidence of water in their lungs, even if alert and fully rational, should be medically examined. Thus, fully conscious survivors who may have aspirated water and are coughing, having trouble "getting enough air into their lungs," or experiencing chest pain should seek medical advice as soon as possible.

▸ Prevention through the use of good safety practices and equipment is the best cure.

6

Hypothermia

"**A** U.S. NATIONAL TRANSPORTATION SAFETY BOARD (NTSB) investigation of an accident in which 3 people, on a small passenger boat
named *El Toro II,* died of hypothermia in 12-degree Celsius (54-degree
Fahrenheit) water after abandoning ship in Chesapeake Bay in December 1993 found that they had, along with 16 survivors, been hanging
on to the side of a rigid buoyant apparatus for 55 to 80 minutes in six-
to eight-foot waves. With time, the three who ultimately died became
fatigued, let go of the raft, and drifted away. The Coast Guard requirement was for the *El Toro II* to carry safety apparatus for only 30 percent
of the passengers. Its single open flotation device was licensed for 20,
although she was complemented to carry 49. In any event, they were
lucky that only 3 died. Had the Coast Guard's 41-foot cutter not arrived
fairly quickly on the scene, with USCG reservists tripling the usual
manpower on board, the rescue would have taken more time and more
deaths could have occurred."

USA Today

Many people recognize the threat to body temperature posed by immersion in cold water. Therefore, some authorities identify specialized protective clothing that personnel at high risk of immersion (e.g., air crew)
must wear. Specifications are also made for the provision of "out of the
water" lifesaving appliances for crews and passengers of boats and ships.
The difficulty arises in deciding the cut-off sea temperature at which such
regulations should come into force. Most authorities opt for a water

temperature of 15 degrees Celsius (59 degrees Fahrenheit), assuming that at that temperature a person of normal build and physical fitness, wearing normal clothing and a life jacket, could survive for six to eight hours. Such an assumption is erroneous. Although some may survive that long in favorable circumstances, many factors can reduce survival time, in some cases to minutes, as we have already seen in chapter 4. In this chapter and the next, we will examine these factors.

Water, unlike air, provides practically no insulation at the skin-water interface. Therefore, heat transfer from the naked body to the adjacent cooler water is extremely rapid. Skin temperature quickly approximates water temperature, increasing the thermal gradient within the body between deep and superficial tissues, down which heat will flow. As already explained in chapter 2, the rate at which this heat flows will depend on the temperature differential; this can be substantial given the physical properties of water. Should the water adjacent to the skin be moving, either because of current or body movement (e.g., swimming), the convective heat-transfer component will increase. Thus, naked humans cool two to five times more quickly in cold water than in air at the same temperature. The rate of heat lost to water, however, will vary significantly between individuals, influenced by a number of intrinsic and extrinsic factors, shown in table 2.2 (page 38).

Everyday clothing, even though quickly saturated with water on immersion, will still provide some insulation. Convective currents are less likely to affect the warmed water molecules retained within the clothing. Conductive heat transfer, however, continues between these molecules and the surrounding cold water. Thus, although a thermal gradient remains between the skin temperature and that of the deep body tissues, it is somewhat smaller than when naked. Therefore, contrary to the advice given some years ago, it is advisable to keep clothing on when immersed in cold water.

Because of the physical properties of water, particularly the lack of insulation at the skin-water interface, thermoneutral water temperature for humans is close to deep body temperature and averages about 35.5 degrees Celsius (96 degrees Fahrenheit) in most individuals. As a result, immersion in water much below body temperature presents a person with a significant thermal threat. This threat has long been recognized.

Low Temperature Threat

On 13 December 1790 a Liverpool physician, Dr. James Currie, stood with a crowd of fellow citizens watching, helpless and frustrated, as the crew of an American sailing ship stranded on an offshore sandbank near the entrance to Liverpool harbor struggled to survive. The ship was partially capsized, slowly breaking up in the surf, with the crew hanging perilously to the rigging or other vantage points. The sea temperature was just less than 5 degrees Celsius (41 degrees Fahrenheit). With time, some of the crew, no longer able to maintain their precarious hold, fell into the sea and drowned. Others managed to hold on and were eventually rescued as both tide and storm ebbed (Currie 1798).

HISTORICAL

The scene Dr. Currie described was not new to most of the spectators, who had previously witnessed several ships in a similar predicament. What was different on this occasion was the presence of Dr. Currie, a thinking person who began to question the mechanisms of the men's condition in such situations. Some of the survivors' accounts of the terrible cold and how it insidiously caused loss of consciousness in those who subsequently died prompted Currie to undertake some experiments involving the immersion of human volunteers in both fresh water and salt water at about 7 degrees Celsius (45 degrees Fahrenheit) (Currie 1798). These appear to be the first recorded experiments on the physiological effects of cold-water immersion in humans. Currie noted how immersion cooled the body much faster than exposure to air at a similar temperature. He observed that body temperature (measured orally) rose initially, for about nine minutes, before it began to fall linearly for the remainder of the exposure.

Currie also observed that when a person exited the water, body temperature continued to fall a further 4.5 degrees Celsius (8 degrees Fahrenheit) before it began to increase and return to normal. This continued fall in body temperature after removal from cold water is now termed the *afterdrop*, a phenomenon that was to assume great notoriety in the infamous Dachau[1] concentration camp experiments by Nazi researchers during World War II. We will discuss the afterdrop in more detail in

---

1. The authors are reluctant to make reference to these unethical experiments on the principle that science should never acknowledge unethical research, for reasons that are too numerous to go into here. But these horrendous experiments did occur, and many people lost their lives in the process. Ignoring the fact that they happened will not bring the victims back to life, whereas acknowledging them will at least serve as a memorial and ensure that they did not die in vain. By not mentioning the names of the perpetrators who conducted the experiments, we feel that they will achieve no "fame," while acknowledging that such experiments occurred should bring condemnation to the political regime under whose authority they were carried out.

chapter 11 but mention it here in conjunction with Currie because he appears to have been the first to report it.

Currie repeated his experiment on a number of occasions, trying different methods of rewarming his subjects. These included bladders filled with hot water applied to the pit of the abdomen, preheated blankets, and immersion in hot water. He found the latter to be the most effective and was surprised both by the speed of recovery it produced and, in particular, by how the subject "was remarkably cheerful and alert the whole evening" after his ordeal!

Currie was not the first to become aware of the significant relationship between water temperature and survival; the opening quotation to chapter 1 of this book, from Herodotus, shows that such knowledge existed in 450 B.C. and probably even before then. In the mid-18th century further testimony to the recognition of this problem appeared in the writings of the British naval physician Sir James Lind (1762)[2]. Both Lind and Currie were obviously aware that cold was associated with body cooling, muscular fatigue, impairment of consciousness, and subsequent drowning. Both also realized the importance of rewarming to reverse the adverse effects of body cooling. Regrettably, successive generations ignored such knowledge, as they did evidence in other areas of history, thereby necessitating its rediscovery at great cost.

Gilley (1850) gives an excellent account of shipwrecks in the Royal Navy in the years 1793–1849, covering the period of the British naval blockade of Napoleonic Europe. Excluding ships lost through enemy action, the Royal Navy lost 410 ships and over 16,000 lives through shipwreck in that period. His accounts of some of those tragedies leave little doubt about the role played by cold in the resulting loss of life.

After Napoleon's disastrous Russian campaign, when he lost thousands of troops from cold, French scientists showed considerable interest in the physiological effects of cooling on animals. These experiments showed a relationship between lowered body temperature, consciousness, and other physiological functions.

In Germany a police surgeon (Reinke 1875), reporting on cases of drunken sailors rescued from canals in winter in Hamburg, recorded the survival of an individual with a rectal temperature of 24 degrees Celsius (75 degrees Fahrenheit). This was the lowest rectal temperature recorded in a survivor up to that time. Others with higher temperatures (e.g., 26.4, 27.7, and 28.4 degrees Celsius; 80, 82, and 83 degrees Fahren-

2. Sir James Lind (1716–1794), a British navy physician claimed by many to be the father of naval medicine, was renowned for conducting what appears to be the first controlled, properly designed (crossed-over) study in dietetic medicine on record, in 1747. He proved that lemons and oranges— already used by the British and Dutch East Indies fleets—were the best cure for scurvy; a finding that helped make the subsequent naval blockade of Napoleonic Europe possible.

heit) did not survive. Before that a Swedish physician (Naucler 1757) had also implicated alcohol in another remarkable account of survival from profound hypothermia in 1757. He described chancing upon a village whose residents were preparing for the funeral of a peasant who had been found "frozen stiff" lying in the snow, having been outdoors overnight. His limbs were stiff and immovable, and his eyes fixed open. Despite the absence of any signs of life, Naucler thought he detected some warmth in the pit of the man's stomach and recommended that the peasant be actively rewarmed with hot fomentations applied to the trunk. Evidently, the man recovered, although the absence of a recorded temperature makes it difficult to make comparisons with other such remarkable cases.

For many years, the lowest recorded body temperature with subsequent survival lay with the remarkable case described by Laufman (1951) of a 23-year-old female vagrant found sleeping outdoors in Chicago, at an ambient temperature of –24 degrees Celsius (–11 degrees Fahrenheit). Although she was apparently dead, on arrival in the hospital her heart was found to be slowly beating. Her deep body temperature was 18 degrees Celsius (64 degrees Fahrenheit). She survived but with severe cold injuries to her hands and feet. Before this incident it was assumed that a body temperature of 24 degrees Celsius (75 degrees Fahrenheit), recorded by Reinke (1875), was the lowest compatible with survival. None of the unfortunate victims of the horrific experiments conducted by the Nazis in the Dachau concentration camp survived deep body temperatures below that level (Alexander 1945). In recent years, however, it has been found that some people, particularly small children, have been successfully resuscitated from much lower body temperatures after accidental cooling in ice-cold water (see chapter 5).

In the late 1930s and 1940s, deep body temperature was deliberately lowered in selected patients, under general anesthesia, as a therapeutic measure for some medical conditions. Subsequently, and indeed in some centers today, induced hypothermia is still used as an adjunct to cardiac and brain surgery. Hypothermia protects the brain from a shortage of oxygen while the circulation is temporarily impaired. At a brain temperature of 28 degrees Celsius (82 degrees Fahrenheit), hypoxic (shortage of oxygen) survival time doubles. Below 22 degrees Celsius (72 degrees Fahrenheit), brain-cell activity (metabolism) ceases and brain survival is much extended. Nowadays, cardiopulmonary bypass has replaced induced profound hypothermia in such surgery.

The use of hypothermia as an adjunct to clinical treatment yielded much knowledge about its physiological effects. Because hypothermia was induced under general anesthesia, however, the observed physiological responses would not necessarily reflect those seen in survival victims. The narcotic effects of anesthetics suppress the violent shivering seen in conscious survival victims, and with it many of its associated metabolic, cardiac, and respiratory physiological responses. Therefore, we must use caution in transferring clinical experience to the survival situation.

ACCIDENTAL HYPOTHERMIA

During a particularly severe winter in the United Kingdom in the early 1960s, it was recognized that significant numbers of elderly people living in poorly heated houses were victims of hypothermia. At that time few clinical thermometers registered temperatures below 35 degrees Celsius (95 degrees Fahrenheit), and hence many of these victims were not correctly diagnosed. Steps were immediately taken to ensure that clinical thermometers were appropriately calibrated, and the British Medical Association (BMA) selected an arbitrary deep body temperature of 35 degrees Celsius (95 degrees Fahrenheit) below which the state of hypothermia was said to exist (BMA 1964). Most now accept this temperature as defining the onset of clinical hypothermia.

The term *accidental hypothermia* describes the condition of patients who are cold simply because the prevailing environmental conditions have overwhelmed their body's ability to remain in thermal balance. The term *secondary hypothermia* applies to patients whose hypothermia is the result of an impaired thermoregulatory system (due to injury, illness, drugs, etc.). Nowadays, people tend to use the term *hypothermia* loosely to describe anyone who is cold and shivering, regardless of whether the deep body temperature is above or below 35 degrees Celsius (95 degrees Fahrenheit).

Acute and Chronic Accidental Hypothermia

The rate of onset of hypothermia will also influence the condition of the victim. A distinction is made between rapid-onset ("acute") and slow-onset ("chronic") hypothermia. The latter may occur in survivors in open survival craft or in those immersed in moderately cold water (20 to 28 degrees Celsius; 68 to 82 degrees Fahrenheit). In such circumstances the survivor may spend many hours attempting to defend body temperature with a consequent depletion of energy stores and loss of body fluid. In acute hypothermia, as occurs classically in cold water (less than 15 degrees Celsius; 59 degrees Fahrenheit), such consequences are less marked. A distinction is made between acute and chronic hypothermia because it may have implications for methods of treatment. In the chronic victim, the patient may be exhausted and have insufficient fluid reserves to sustain normal circulation on rewarming. With the acute victim, in the absence of other complications such as near drowning, the problem is associated with lowered body temperature. Once the body warms up, the physiological regulatory processes of the body are able to continue as normal.

Diagnosis: Signs and Symptoms of Hypothermia

In the absence of a thermometer it is difficult to differentiate between someone who is just cold and one who is suffering from the early stages of hypothermia. In practice it makes little difference because the initial treatment is similar. One may obtain an indication of the severity of cooling by feeling the temperature of the casualty's deep armpit; that of the truly hypothermic victim feels particularly cold ("like cold marble") when compared with someone who is just superficially cold.

A Dangerous Effect of Hypothermia

"I remember sitting in the cockpit and noticing one of the buttons of my oilskin jacket was undone. For some reason I was unable, and unwilling, to do anything about it, although I knew I should. But one of the effects of hypothermia is that your brain just seems to come to a grinding halt, which of course makes things worse" (M. Sheahan, Fastnet race survivor from the yacht *Grimalkin*).

The signs and symptoms of hypothermia are primarily the consequence of the inhibitory effects of reduced temperature on cellular (metabolic) activity in most of the body organs, but in particular the brain. Many of the outward effects of hypothermia, however, are the result of the associated peripheral muscle and nerve cooling, which usually precede brain cooling.

Overtly, victims are invariably shivering, and if conscious, their speech tends to be slurred. However, rarely are they particularly communicative (they tend to become introverted) other than possibly complaining of cold. They may also exhibit uncharacteristic behavior or personality. They will usually be uncoordinated with a general slowing in physical and mental activity. This condition will increase the incidence of errors of omission or commission and, in turn, may lead to poor judgment, bad decisions, reduced perception, or dropping or damaging vital equipment. In general, hypothermic individuals will be performing far below par and be a risk both to themselves and others. With continued cooling, victims will become progressively more withdrawn, consciousness will deteriorate, and with it other body functions, until eventually victims become unconscious and die (figure 6.1).

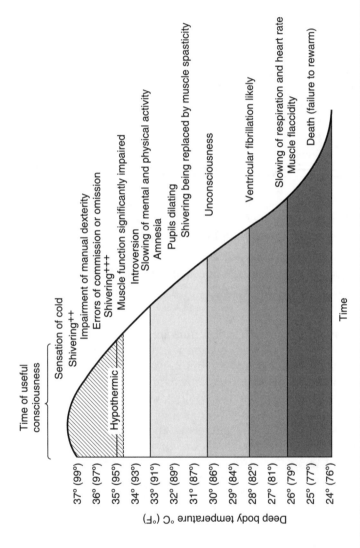

Figure 6.1 Empirical curve showing the fall of body temperature against time with corresponding signs and symptoms experienced at each temperature. Adapted from Golden 1973.

In profound, or deep, hypothermia, the patient will be deeply unconscious and will not react to painful stimuli. The reduced oxygen requirement of very cold cells and organs, which are almost in a state of suspended animation, results in a heart rate and breathing that may be difficult to detect. The overall appearance may be one of death. The implications for the immersion casualty and the life raft "survivor" will be discussed later, but first we describe the more important signs of impending hypothermia; recognition of these should provoke preventive action.

Shivering

Strictly speaking, shivering is not a reliable sign of hypothermia. Despite this, the presence or absence of shivering has been used as an indicator that body temperature is under threat or has reached dangerously low levels. This contradiction results, in part, from the misuse of data from the clinical literature noted earlier, as well as from the wide variation of circumstances and conditions that can result in hypothermia.

Immersion in cold water results in a sudden lowering of skin temperature. The consequent dynamic response and high firing rate of the cutaneous cold receptors initiate shivering. Paradoxically, during this early stage of cold exposure, deep body temperature may actually rise (reported by James Currie in 1798) because heat production from shivering occurs simultaneously with peripheral vasoconstriction and before the establishment of a thermal gradient for heat loss from deep body tissues. At this time the presence of shivering is not a sign of lowered deep body temperature.

In a naked individual, skin temperature falls and begins to stabilize close to water temperature in 5 to 10 minutes (figure 6.2). When this happens, the rate of firing of the cutaneous cold receptors and associated shivering declines. As a result of the stabilization of skin temperature and decline in shivering, the immersed individual may feel surprisingly comfortable at this time.

After a period that varies with factors such as body insulation and water temperature but averages about 15 minutes, deep body temperature will start to fall below normal levels as a thermal gradient is established between it and the water surrounding the body. Deep (central) temperature receptors detect this fall and initiate the return of shivering. In humans a fall in deep body temperature is a more powerful stimulus to shivering than a corresponding fall in skin temperature. If core temperature continues to fall, shivering intensity increases. Initially it comes in intermittent bursts of mild to moderate intensity. The intermittent muscle tremor associated with shivering will interfere with manual dexterity, making tasks requiring fine movement difficult. Furthermore, because both the protagonist and antagonist (flexor and extensor) muscles

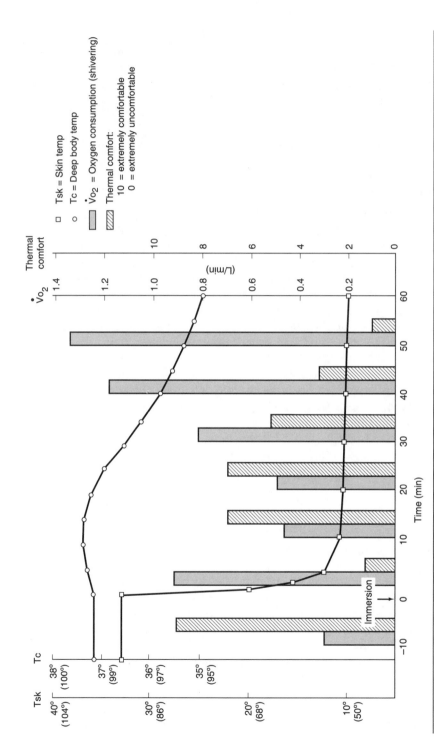

Figure 6.2 Changes in skin and deep body temperatures with time during immersion in cold water and associated shivering and thermal-comfort responses.

are contracting simultaneously in shivering, controlled coordinated movement becomes difficult and the limbs feel stiff. Should the cold exposure be of such severity that it numbs the sensation of the fingers, further impairment to manual dexterity occurs and escalates with time. These factors should be considered when designing or selecting lifesaving appliances. Equipment that relies for its activation and use on fine movement or a good sense of touch is of little help in a survival situation in a cold environment.

With continued deep body cooling, intermittent shivering progresses to continuous shivering in the major muscle groups of the limbs. Its intensity gradually increases until the limbs may shake violently. At this stage the individual will be shivering maximally at an intensity equivalent to about 46 percent of maximum aerobic capacity (Golden et al. 1979). The magnitude of the shivering response is proportional to the level of the threat: A gradual fall in deep body temperature will elicit a slow but gradual increase in shivering, whereas a rapid change in body temperature will trigger vigorous shivering.

At this stage shivering is a sign, in most individuals, that body temperature is under threat. But the absence of shivering cannot be interpreted as indicating that the body is not under threat; shivering may be absent in unconscious or intoxicated hypothermic victims because of a central inhibition of the shivering response. The intensity of the shivering response will be significantly smaller in people who have a low blood-sugar level (chapter 8). In some cases, shivering may be completely absent even with a deep body temperature in the region of 35 degrees Celsius (95 degrees Fahrenheit) when the blood-sugar level is at 50 percent of normal values (Passias et al. 1996). This situation may occur in life-raft survivors on minimal rations. It may also occur in those who have not eaten but have consumed as little as two pints of standard-strength beer and have exercised moderately for about 20 minutes. Alcohol impairs the ability of the liver to maintain blood-sugar levels (Haight and Keatinge 1973). Consequently, in cold environments, such people are at increased risk of developing hypothermia. Furthermore, besides not having a normal shivering response, they perceive less discomfort from the cold and are less aware of the danger they are in. It is probably this reaction that has led to the myth that alcohol "keeps the cold out!"

If body temperature stabilizes, shivering becomes less intense and intermittent. It will remain like this until the fuels to support it are exhausted. In this situation, shivering is a sign of a lowered, but not falling, deep body temperature.

Some accounts in the literature claim that shivering declines in intensity below a deep body temperature of 35 degrees Celsius (95 degrees Fahrenheit). Although, as noted above, this might be true in exhausted, chronically hypothermic casualties or in drug-, alcohol-, or anesthetic-induced hypothermia, it is not necessarily the case in acute hypothermia. With rapid cooling, victims with low body temperatures (less than 35 degrees Celsius; 95 degrees Fahrenheit) may still shiver violently. Below a deep body temperature of 30 degrees Celsius (86 degrees Fahrenheit), they are likely to be at the spastic stage of muscle activity. Irrespective of actual body temperature, the magnitude of the shivering response will determine the respiratory and heart rates observed. Classically, these are depressed in hypothermia, but they will be raised while shivering is present.

If deep body temperature continues to fall, violent intermittent shivering will gradually be superseded by progressive development of muscle spasm. The limbs will develop a fixed, flexed attitude, causing the body to adopt a fetal, almost spastic, position. At this stage the survivor is virtually incapable of assisting with his or her rescue. Eventually, depression of muscle-cell activity occurs, and death soon follows from hypothermic cardiac arrest.

Skin

Accounts of hypothermic victims by different authors describe the skin as being either pale or bright red, with a blue tinge to the lips and nail beds. Such differences are possible, depending on the circumstances in which the individuals were cooled and the stage of their cooling or recovery when examined. Under normal circumstances a cold stimulus results in immediate vasoconstriction, producing pallor. In conditions where deep body temperature is near normal but the tissues beneath the skin cool to below 12 degrees Celsius (54 degrees Fahrenheit), the muscle in the walls of skin blood vessels becomes paralyzed by cold and relaxes (Keatinge 1969). Blood from deeper unaffected blood vessels floods into the more superficial relaxed ones. This is termed *cold-induced vasodilatation* (CIVD). This filling of the microscopic capillaries in the skin gives it a flushed brick-red appearance. The red coloration, due to the nearly 100 percent oxygen content of the blood, is the result of both minimal demand for oxygen by the cold tissues of the skin and a reduced ability of blood to release oxygen when cooled.

A common localized example of CIVD is the warm sensation experienced in bare hands after a period of contact with snow. This circumstance will occur only if the remainder of the body is warm. People who fillet frozen or chilled fish are familiar with, and indeed reliant on, this phenomenon to assist with their manual dexterity.

When the remainder of the body is cold (e.g., during cold immersion) and deep body temperature is being threatened, deeper blood vessels remain constricted. Thus, although the most superficial small vessels may be open because of cold paralysis, with resulting reddening of the skin, little or no effective flow occurs between deep and superficial vessels. In this situation the reddening of the skin is often mistaken for CIVD but, in fact, the red color is simply blood trapped in the skin unable, because of its low temperature, to release its oxygen. The distinction is shown by the time it takes for a white pressure point created by gentle finger pressure on a patch of red skin to dissipate and refill with blood. After removal from the cold water, the skin will do one of the following two things:

1. Return to a pale coloration when the temperature of the cold-paralyzed vessels returns to above 12 degrees Celsius (54 degrees Fahrenheit) and normal vascular muscle contractility is restored. Blood contained in the skin is then squeezed back into deeper circulation.

2. Continue to remain red with discreet blue patches.This occurs if the temperature of the blood trapped in the skin increases sufficiently to release its oxygen and turn blue before normal blood flow returns.

Severe Hypothermia

With the continued progressive decline in deep body temperature, the victim becomes more introverted as cerebral activity declines (see figure 6.1 on page 102). When brain temperature is below 34 degrees Celsius (93 degrees Fahrenheit), the period of useful consciousness is limited as consciousness gradually becomes impaired. Amnesia (loss of memory) may also occur about this time. Consciousness is eventually lost at about 30 degrees Celsius (86 degrees Fahrenheit) in acute hypothermia but may still be present at this temperature in chronic hypothermia. At this stage, the pupils of the eyes also dilate. In secondary hypothermia the absence of shivering, marked slowing of respiration and heart rate, and difficulty of detecting a peripheral pulse frequently make the diagnosis of death difficult outside a hospital environment and poses some worrisome management problems. Below a cardiac temperature of 28 degrees Celsius (82 degrees Fahrenheit), the heart may suddenly and spontaneously arrest either through *ventricular fibrillation* (V.F.) or later in relaxation *(asystole)* when the cardiac muscle itself becomes paralyzed by cold. V.F. may result from rough handling at temperatures around 28 degrees Celsius (82 degrees Fahrenheit). Those involved in the rescue of hypothermic casualties must appreciate this and act accordingly (see chapter 11).

Clearly, therefore, with cooling and the associated progressive loss of consciousness, the requirement to protect the airway of the immersed victim is of paramount importance.

LIFE JACKETS AND FLOTATION DEVICES

Life jackets, or personal flotation devices (PFDs) as they are called in the United States, are intended to provide adequate buoyancy to support an unconscious clothed person in water while maintaining the airway clear of the surface. The amount of buoyancy specified to meet these criteria varies between nations, with a resulting conflict in standards. In recent years, many of the previous standards have been amended and currently approximate each other at around 150 newtons (N), about 34 pounds of buoyancy. Minor differences still exist between countries. Some nations recognize that particular activities require different levels of buoyancy, varying from above 275 newtons (62 pounds) down to about 50 newtons (11 pounds). The upper level is required for high-risk groups wearing special protective clothing, which alters the normal flotation angle (see discussion on page 132), or for those likely to be carrying heavy equipment (e.g., military personnel). The lower end of the scale is intended for those who can swim and are deemed to be at low risk (e.g., those in relatively warm waters operating close to shore). They are also designed for water sports requiring unrestricted movement, such as canoeing and windsurfing. The low-buoyancy appliances are usually termed *buoyancy aids,* although terminology may differ between countries. To qualify for "life jacket" classification in the United Kingdom, a device for adults must have in excess of 150 newtons (34 pounds), whereas a "buoyancy aid" needs only 50 newtons (11 pounds). A book by Brooks (1995) specifically about life jackets contains a full discussion on the various national standards.

Generally, buoyancy aids are intended to provide additional buoyancy to conscious people to help them remain afloat for a short interval until they are rescued or can swim to a place of safety. They are meant for people who can swim, are confident in the water, and are within easy reach of rescue. Such devices provide insufficient buoyancy to help someone suffering from cold shock, especially in turbulent water. They will certainly be inadequate for a survivor in the early stages of hypothermia, when consciousness is declining and muscle coordination is impaired. In contrast, life jackets should have sufficient and appropriately distributed buoyancy to bring the wearer to the surface, provide self-righting, and keep the airway clear of the water. To do this, life jackets must have a minimum of 150 newtons of buoyancy. To guarantee safety, the life jacket must also be well designed and correctly secured to the body.

Wave Splash

In practice, life jackets can fail to keep the mouth clear of the water if not correctly adjusted to ensure a secure fit. Ideally, they should have a crotch

The time to learn how to secure a life jacket is not just prior to abandonment.
Photograph courtesy of Frank Golden and Michael Tipton.

strap to prevent them from riding up over the shoulders, a fact that was learned at great cost during World War II (Llano 1955). Sadly, because there are no regulatory requirements for crotch straps, many life jackets on the market do not have them. Even correctly secured life jackets are likely to be effective only if the sea is relatively calm. Because life-jacket buoyancy is mainly distributed over the front of the upper torso to facilitate self-righting, and the dependent legs act as a sea anchor, waves apply a turning moment to the body. The result is that within one or two waves, a totally relaxed or unconscious person will turn to float facing the oncoming waves.

Sea Canoeing Disaster

In March 1993 a party of eight 16- and 17-year-old school children, accompanied by a schoolteacher and two instructors, were sea canoeing as part of an outward-bound training course off the south of England. Apart from the instructors, all were novice canoeists. The teacher was experiencing

(continued)

(continued)

great difficulty in remaining upright from the outset, so the senior instructor remained close to him to provide personal supervision. Consequently, both became separated from the remainder of the party. All the group were wearing neoprene wet suits and flotation devices that had an inherent 60 newtons of fixed buoyancy but were capable of being orally inflated to provide an additional 100 newtons or more of buoyancy. Tragically, none of the children had been instructed of this potential, so when they later found themselves in the very cold water, after repeated capsizings in one-meter (three-foot) seas, they had insufficient buoyancy to keep their airways clear of the water.

"Once we were all in the water and were too tired to attempt any further attempts at trying to bail and board the canoes, we abandoned them and linked arms to form a long line. The swell was repeatedly breaking over our heads. Later we tried to form a circle but were unable to do so. All of us felt cold and were shivering almost immediately after water entry. Sometime later one girl became withdrawn and lapsed into unconsciousness. Subsequently she was experiencing difficulty in breathing and started foaming around the mouth. Soon, one of the boys experienced a similar problem, profound confusion, talking rubbish, followed shortly after by unconsciousness with copious bubbles around the mouth. Gradually most of the remainder began to experience similar problems before the rescue helicopters arrived on scene."

The instructor and four of the eight students were rescued alive. An autopsy found that the remaining four students died from drowning.

Should the waves be small with high frequency, they may contain insufficient energy to overcome the inertia of a body wearing soaked clothing. Water will then well up around the neck and mouth of the wearer with each wave. Alternatively, water will simply slap against the leading edge of the life jacket and splash over the face. This occurrence is more likely when the mouth-to-water distance is small, as may be the case with those wearing buoyancy aids or poorly designed, badly fitting, or inadequately inflated life jackets (see preceding anecdote). Some twin-lobed life jackets direct water toward the mouth in the channel formed by the lobes. Small waves running against a current or tidal stream can produce a steep wave face that can be particularly troublesome in this respect. In larger breaking waves the narrow crest of the wave, or white aerated water, may periodically wash entirely over the face of the wearer. This phenomenon is termed *wave splash* (figure 6.3).

Figure 6.3   Wave splash—the breaking wave over the face of a person wearing a life jacket.

By interfering with normal breathing, the repetitive splashing of waves on the face quickly becomes intolerable. A conscious person can detect an approaching wave and voluntarily synchronize breathing to avoid inspiring water as the wave washes over. However, it quickly becomes tiresome and annoying to have to do this continually. Furthermore, because the frequency of the waves is unlikely to coincide with natural respiratory frequency, breathing becomes a conscious act and thus uncomfortable. Consequently, people in such situations quickly learn to turn their backs to the waves; there is no respiratory discomfort experienced with the wave breaking over the back of the head because you will subconsciously cease inspiration as the wave hits the head. Keeping the back to the waves, however, requires a paddling movement by the arms to counter the turning moment applied to the torso by the waves. Such arm activity can accelerate heat loss. It also requires a level of consciousness and muscular coordination, both of which will be impaired with prolonged cooling. Hence, it is likely in such sea states that the "survivor" will drown before severe hypothermia develops and certainly before death from hypothermia occurs.

To counter the threat from wave splash, the British navy designed, tested, and introduced the "splashguard" in the early 1970s (figure 6.4). These are now fitted to all standard Royal Navy life jackets and to those used by the German navy (Hermann and Stormer 1985).

Some commercial varieties (sometimes termed *spray hoods*) are available but, because comprehensive international standards do not exist at this time, the materials from which they are manufactured and their

Figure 6.4   Picture of a Royal Navy splashguard.

design vary considerably, often not for the better. Some poorly designed splashguards have given the concept a bad name in some quarters. As with many forms of protective equipment, however, a distinction should be made between, on the one hand, an excellent concept with a poor design and, on the other hand, a poor concept. The splashguard is undoubtedly an important lifesaving concept; the secret is to buy one that works.

A well-designed splashguard

- ▶ is made from a clear plastic that is not too pliable (this will prevent it from collapsing onto the face under pressure from the wave);
- ▶ contains a suitably positioned vent to permit the escape of the carbon dioxide from the expired air and its replacement with fresh air, without facilitating flooding beneath the hood with each passing wave (i.e., complies with a European standard); and
- ▶ is easily donned and doffed.

Detractors of splashguards claim that they reduce visibility because the clear visor quickly clouds over with condensation from expired air.

Although this is true, a conscious survivor can periodically wipe the visor clear as necessary, whereas the semiconscious or unconscious survivor is not interested in the view! For such individuals, the splashguard will extend survival time from the point at which they begin to lose consciousness, to the point at which they reach lethal deep body temperature. This extension of survival time can be significant (figure 6.5).

How Likely Are Immersed Victims to Die From Hypothermia?

In spite of the great capacity of cold water to extract heat from the immersed body, hypothermia, unlike cold shock, is unlikely to be a problem within 30 minutes of head-out immersion for a fit, clothed adult, even in water as cold as 5 degrees Celsius (41 degrees Fahrenheit). In exceptional circumstances, however, rapid body cooling may occur when a person aspirates a large volume of water (see chapter 5). Drowning will then pose a more serious threat than hypothermia. More usually, although normal thermoregulatory responses may help slow the rate of heat loss, they will not be able to match it and body temperature will continue to fall. Thus hypothermia will eventually occur, although its rate of onset will vary between individuals for the reasons explained in chapters 2 and 7.

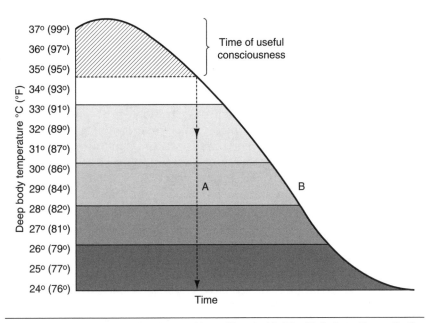

Figure 6.5   Theoretical survival times, without (A) and with (B) a life jacket with an effective splashguard.

As a rule, during head-out immersion in laboratory conditions, the deep body temperature of an average young male adult wearing nonspecialized clothing (1 clo in air and 0.06 clo in water) will fall 2 degrees Celsius (3.6 degrees Fahrenheit) to about 35 degrees Celsius (95 degrees Fahrenheit) after approximately one hour in stirred water at 5 degrees Celsius (41 degrees Fahrenheit). In water at 10 degrees Celsius (50 degrees Fahrenheit) the same decline in body temperature will occur in two hours; and in water at 15 degrees Celsius (59 degrees Fahrenheit) in three to six hours. A deep body temperature of about 35 degrees Celsius (95 degrees Fahrenheit) is associated with the onset of decline in useful consciousness. Thereafter, physical and mental capability progressively deteriorate, and individuals become less able to help themselves. This deterioration will occur well before loss of consciousness from hypothermia.

In contrast to the laboratory situation, in open water, because of the rich supply of blood vessels to the scalp that do not cold vasoconstrict, heat loss from the unprotected head through forced convection and evaporation may increase. This will increase the rate of cooling and hasten the onset of loss of useful consciousness. Thus, if the head is partially immersed body cooling will be significantly faster than the rates outlined earlier. Therefore, perhaps it is not surprising to find that the 50 percent survival time for a clothed person in open water is about the time it takes an experimental subject to cool to 35 degrees Celsius (95 degrees Fahrenheit).

PROTECTION AGAINST HYPOTHERMIA

Protection against hypothermia follows logically from the basic principles of heat conservation already outlined in chapters 2 and 3. Convective cooling can be reduced by remaining as motionless as possible in the water while adopting a position to reduce heat loss. The influence of exercise and posture on survival time is explained in the next chapter.

Ideally a person should get out of the water as soon as possible. In this respect it is better to float on a wooden plank, inverted hull, or open boat than remain immersed (Steinman et al. 1987). Although being out of the water may feel colder than being in it, the rate of heat extraction from the body while immersed will far exceed that in air. Better still, a person should board a life raft or other suitable craft where some shelter from the ambient conditions can be found and efforts made to try to control the surrounding environment (discussed in chapter 9).

Because you cannot always board a raft dry, some form of personal protective equipment is essential. In chapter 4, we discussed the merit of a simple uninsulated waterproof outergarment in protecting against cold shock. Such a device will convey some added protection against hypothermia provided the garments worn beneath it remain dry. Better protection against hypothermia can be achieved if these undergarments are made from a hydrophobic material, are resistant to compression, and include several layers. Because little heat is lost from cold, vaso-constricted, unexercised limbs, and because the major route for heat loss is from the torso and head, the undergarment should cover these areas and the upper third of the limbs.

Besides appropriately insulated undergarments, you should wear a good specialized waterproof immersion survival suit, with gloves and appropriate head covering to reduce evaporative and convective heat loss from the scalp. A good quality life jacket compatible with the suit is also necessary. Those designing or buying immersion-protective equip-ment should regard the combination of immersion suits and life jackets as an integrated survival system (chapter 7).

Even when out of the water and after gaining the sanctuary of a life craft, the threat from hypothermia is far from over. This problem is dis-cussed in chapter 9. Chapter 11 deals with the immediate care and man-agement of hypothermic casualties.

CONCLUSIONS

"People helped each other, but some, wearing underwear only, became quite apathetic."

MV Estonia, Final Report 1997

Hypothermia will constitute a threat to life for those wearing life jackets in relatively calm water and those with life jackets fitted with splashguards (spray hoods) in rougher water. Unless the sea is calm, those wearing life jackets without splashguards will probably die before life-threatening hypothermia develops. Peripheral muscle cooling and dulled brain ac-tivity during the early stages of hypothermia can lead to a loss of useful consciousness and drowning. As explained in chapters 2 and 7, people will cool at different rates for a variety of reasons. Those who cool more quickly are likely to drown first, whereas those who have better insula-tion should survive longer.

It is concluded that body cooling and the onset of hypothermia will increase the possibility of water aspiration and drowning before death from hypothermia. The probable sequence of events and the nature of the threat for the survivor immersed in open water is as follows:

1. Cold shock—possibly leading to drowning or cardiovascular prob-lems

2. Muscle and peripheral nerve cooling—possibly leading to incapacitation and drowning

3. Hypothermia—possibly leading to incapacitation and drowning

4. Hypothermia—possibly leading to cardiac arrest

In all but tropical waters, hypothermia as a cause of death is more likely to occur in life craft survivors than those in the water, who are more likely to drown (see chapter 9). Drowning victims who have aspirated very cold water may have an associated severe level of hypothermia (see chapter 5).

Chapter Summary and Recommendations

▸ Although not initially recognized as a threat, since World War II many have regarded hypothermia as the primary threat to be faced on immersion in cold water.

▸ While still constituting a threat, death from hypothermia is now considered less of a threat than drowning, although it may be a contributory factor.

▸ The loss of useful consciousness associated with early stages of hypothermia will lead to incapacitation. This may result in drowning, especially in rough water and when a person does not wear a life jacket and splashguard.

▸ Hypothermia remains a potent threat to those in open boats and life rafts in all but tropical waters.

▸ The signs and symptoms of hypothermia are understood. But the presence or absence of shivering, or blue coloration of the skin, should not be regarded as a reliable sign of hypothermia.

▸ Body cooling during immersion can occur five times more quickly than it does in air at the same temperature. In no circumstances are you better off in the water than out of it.

▸ Ordinary clothing provides some protection against heat loss in water. Even better is a waterproof outer layer, covering the underclothing. Ideally, some of the underclothing should have low compressibility and be manufactured from hydrophobic material to improve insulation.

▸ Wear a life jacket that provides good mouth-to-water clearance, preferably with a splashguard. The life jacket should have good self-righting qualities and be fitted with a crotch strap. Always make sure it is correctly secured before entering the water.

Prevention is the best cure:

a. Avoid becoming immersed in the first place, if possible. Board the survival craft dry.

b. If immersion is unavoidable, don additional clothing with an outer protective layer. Use gloves and head covering to insulate these areas.

c. Make sure that your life jacket has adequate buoyancy and is correctly adjusted.

d. If water entry is necessary, do so slowly and get out as soon as possible.

e. Perform necessary vital actions early, before cold impairs manual dexterity.

f. If you are unable to leave the water, adopt heat-conservation measures by, for example, reducing both movement and surface area in contact with water.

g. In a survival craft, optimize the thermal insulation of your environment as soon as possible (see chapter 9).

7

Survival Time in Cold Water

GENERALLY, A SEARCH ISN'T COMMENCED until a boat is overdue for more than 24 hours. Fortunately for the passengers and crew of the fishing boat *Cougar,* which sank in 13-degree Celsius (55-degrees Fahrenheit) water 56 kilometers (35 miles) off the coast of Oregon in September 1998, the absence of a normally punctilious individual, Captain Liddell, from a meeting raised the alarm and triggered the search. He, along with five others, was rescued after 18 hours in the water. Sadly, four died before rescue. Had the normal search practice been operated, all would have died.

Knowledge of the estimated survival time of an individual immersed in cold water is seminal in the formulation and execution of search and rescue polices, as well as being important in the selection and purchase of protective survival equipment. For example, in the *Cougar* incident the availability of life rafts or immersion suits could have prevented loss of life.

The usefulness of accurate estimations of survival time for predictive and prescriptive purposes has concentrated a good deal of attention on the topic over the last 50 years. In spite of this, however, and in some cases because of it, the prediction of survival time in cold water remains more art than science. Confusion abounds and an accurate, robust, and comprehensive set of predictions remains elusive. As recently as 1996 a U.K. Health and Safety Executive report on the subject concluded that "there is still a need to define a realistic estimate of probable survival times for people immersed in the (North) Sea" (Robertson and Simpson 1995). In the following section we examine why this is the case.

ESTIMATIONS OF SURVIVAL TIME

The first attempt to quantify the precise relationship between water temperature and survival time was made by George Molnar in 1946. He performed a retrospective analysis of an unspecified number of selected U.S. navy records of ship sinkings and aircraft ditchings during World War II. Using data only from those incidents for which precise details of seawater temperature and time of immersion were available, he compiled a graph of survival time against water temperature (figure 7.1). The graph displays a curve above the highest recorded survival times that "represents a limit of tolerance which probably few men can exceed and many cannot even approach."

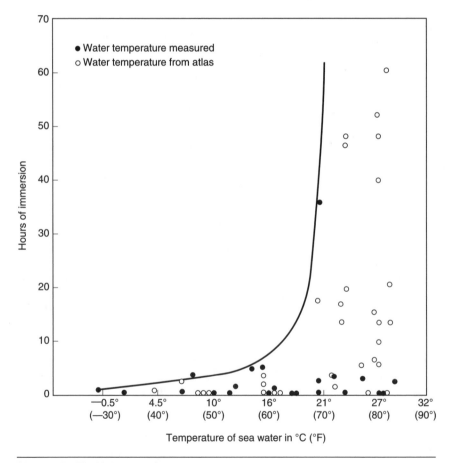

Figure 7.1   The Molnar (survival) curve.

Reprinted from Molnar 1946.

Molnar's analysis was limited to only five data points for water temperature at or below 7 degrees Celsius (42.5 degrees Fahrenheit). For temperatures below 10 degrees Celsius (50 degrees Fahrenheit), he employed survival times extrapolated from the infamous immersion experiments undertaken at the Nazi Dachau concentration camp. Molnar's resulting survival curve is therefore relatively crude because it provides guidance only to the nearest couple of hours. But it did identify, for the first time, that survival time in water below 15 degrees Celsius (59 degrees Fahrenheit) was relatively short and suggested that above this temperature it increased reasonably quickly.

In 1962 Barnett published an empirical predictive survival graph based on Molnar's original (figure 7.2). His expanded six-hour time axis made his curve more user friendly. To overcome the uncertainty of the gray area on the border of Molnar's original single curve, he substituted two curves: one delineating a relative "safe" zone, the other a "lethal, 100 percent expectation of death" zone. The large area between these two curves was labeled "marginal, 50 percent expectation of unconsciousness, which will probably result in drowning."

In 1969, Keatinge proposed guidelines based on the experience of laboratory experiments and immersion incidents. Golden (1976) used both

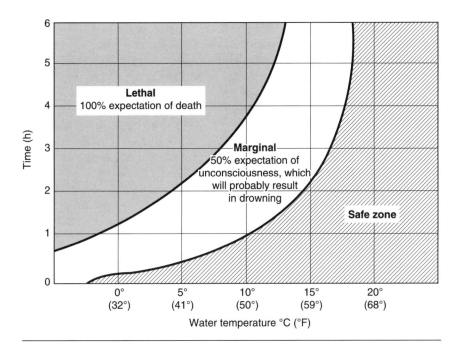

Figure 7.2   The Barnett (survival) curve.

experimental data and case histories of shipwreck and aircraft survivors in producing a curve showing the estimated "50 percent survival times" in cold water (that is, the survival time of the average person).

The most recent contribution to this controversial area comes from the still ongoing U.K. National Immersion Incident Survey (UKNIIS) (Oakley and Pethybridge 1997). This voluntary reporting scheme, begun in 1990, has the aim of validating the various predictive survival curves described earlier. Some of the rescue organizations send the authors information relating to persons rescued, predominantly from the sea.

In 1975, Hayward and his colleagues were among the first to produce a mathematical formula to calculate survival times for cold-water immersion victims. They derived the formula from an analysis of the cooling rates encountered in 15 young, fit, lightly clothed volunteers who were immersed in cold water (5 to 18 degrees Celsius; 41 to 64 degrees Fahrenheit). They mathematically extrapolated the cooling rates to a body temperature at which it was considered death would inevitably occur (30 degrees Celsius; 86 degrees Fahrenheit). By implication, if rescue did not occur before that time, survival would be unlikely.

Subsequently other, and more complex, engineering-based mathematical models have been used to estimate survival time and recommend suitable levels of clothing insulation (Wissler 1981; Hayes and Cohen 1987; Tikuisis et al. 1988). Among them have been mathematical models of the human thermoregulatory system. Figure 7.3 shows the output from one such model. Since this version of the model was produced it has been through several iterations that have altered the predictions somewhat. One of these increased the "lethal" arterial blood temperature from 33 to 34 degrees Celsius (91 to 93 degrees Fahrenheit). Again, such models have been based largely on laboratory studies of young, healthy males, supplemented to a limited degree by information from animal work. The models include "adjustments" to produce predictions that are closer to those observed in real life.

Thus, the two main sources of the information used to estimate survival times in cold water are reviews of actual emergencies and laboratory experimentation supplemented by mathematical manipulation and extrapolation. Whatever the source of the data, the various predictions tend to agree more closely at very cold water temperatures and vary more as water temperature increases (tables 7.1 and 7.2 on page 124). The likely explanation is that the cooling power of the environment at the coldest temperatures overwhelms the various physiological factors that might cause differences among individuals, whereas in warmer water these factors are more likely to be a source of variation among individuals.

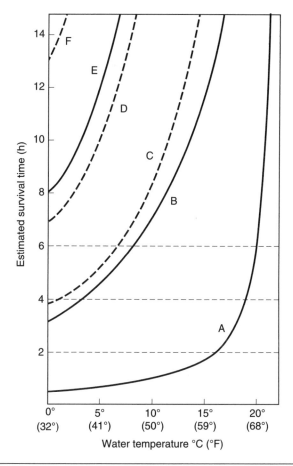

Figure 7.3 Mathematically derived curves showing estimated survival time (time to an arterial temperature of 33 degrees Celsius; 91 degrees Fahrenheit) for people in cold water, indicating the effect of external (clothing, clo) and intrinsic (subcutaneous fat) insulation on body cooling rate. Curve A = 0.06 clo, naked, thin man. Curve B = 0.33 clo, thin man wearing an uninsulated dry suit. Curve C = 0.06 clo, naked, fat man. Curve D = 0.33 clo, fat man wearing an uninsulated dry suit. Curve E = 0.70 clo, thin man wearing an insulated dry suit. Curve F = 0.70 clo, fat man wearing an insulated dry suit.

Adapted from Hayes and Cohen 1987. (© Crown copyright 1987/MOD).

Sources of Error in the Estimation of Survival Time

Several factors make the prediction of survival time in cold water difficult. These include the ambiguity of the definition of survival time, the lack of well-documented data from actual incidents, the consequent reliance on extrapolation of data from innocuous laboratory-based

Table 7.1   50 Percent Survival Times (Hours) for Lightly Clad Males, From Various Authors

Water temperature	Molnar	Hayward	Golden	Tikuisis
5 °C (41 °F)	1	2.2	1	2.2
10 °C (50 °F)	2.2	2.9	2	3.6
15 °C (59 °F)	5.5	4.8	6	7.7

From Oakley and Pethybridge 1997.

Table 7.2   Immersion Times (Hours) Resulting in "Likely Death" for Lightly Clad Males, From Various Authors

Water temperature	Molnar	Keatinge	Nunnely & Wissler	Allan	Lee & Lee
5 °C (41 °F)	2.3	0.9	1.1	1.5	1
10 °C (50 °F)	4		2.6	2.5	3
15 °C (59 °F)		4.5	3	9	7

From Oakley and Pethybridge 1997.

exposures, and individual variability. More important than all of these, however, are the variability in sea state and the exclusion of the possibility of death from drowning (see chapter 5). We consider some of these issues in the section that follows.

Interpretation

One common source of error occurs in interpreting the curve drawn on a graph depicting survival time. As we saw in the previous section, this curve can represent a limit of tolerance, a relative safe zone, a lethal zone, a marginal zone, or estimated 50 percent survival time. Other authors provide information relating to a "good prospect of survival." The 50 percent survival curve has been mistakenly taken to represent the time at which an individual would have a 50-50 chance of survival rather than the survival time of 50 percent of individuals. Furthermore, people often forget that this time theoretically exceeds the survival time of 50 percent of individuals.

The problems just outlined are exacerbated by replication of the various survival curves without the associated explanations and caveats, and by the ill-conceived formulation of hybrid curves.

Lowest Temperature Compatible With Life

The prediction of survival time in cold water requires an *a priori* assumption about the deep body temperature at which death occurs. The temperatures chosen have varied between a deep body temperature of 28 degrees Celsius (82 degrees Fahrenheit) and a "blood (arterial) temperature" of 34 degrees Celsius (91 degrees Fahrenheit). Inherent within this assumption is a second about the cause of death on immersion.

Data obtained by the infamous Nazi researchers at Dachau during World War II suggested that during immersion in calm water under laboratory conditions, death occurs from cardiac causes rather than respiratory causes. The lethal body temperature for people immersed in cold water was reported to be about 24 to 26 degrees Celsius (75 to 79 degrees Fahrenheit), although cardiac arrest from ventricular fibrillation frequently occurred below 28 degrees Celsius (82 degrees Fahrenheit). Concern over the ethical and scientific accuracy of these data casts doubt on their precise veracity, but when taken together with anecdotal accounts from many different sources, we can conclude that most of the reported Dachau data are reasonably accurate. Nevertheless, many individuals have survived much lower deep body temperatures after accidental exposure to cold, the current record being a rectal temperature of 13.7 degrees Celsius (56.6 degrees Fahrenheit) in a young female skier (Gilbert et al. 2000).

Thus, even a cursory glance at the literature shows that a single lethal deep body temperature does not exist and that the temperature thought to be incompatible with life can vary greatly among individuals.

The temperature selected by those wishing to predict survival time is generally higher than 25 degrees Celsius (77 degrees Fahrenheit).This is to allow for the fact that at relatively high deep body temperatures local cooling of superficial muscles and nerves inhibits their function to the point where the victim can manage only limited self-help. Death by drowning can then occur before death from hypothermic cardiac arrest (chapter 6). The deep body temperature and time at which this incapacitation occurs, however, are potentially just as variable as the "lethal" temperature. Therefore, the selection of the lowest body temperature compatible with life remains a major potential source of error in the estimation of survival time.

The fact that most predictions of survival time ignore the initial and short-term responses to immersion (chapter 4) compounds this error. The evidence presented in previous chapters suggests that drowning consequent to cold shock represents the major risk on immersion in

The passengers and crew were forced to abandon the cruise ship *Prisendam* in the frigid waters of the Gulf of Alaska after a fire broke out on board.
Official U.S. Coast Guard photograph.

cold water (less than 15 degrees Celsius; 59 degrees Fahrenheit) and that in very cold water (less than 5 degrees Celsius; 41 degrees Fahrenheit) survival time for many is measured in minutes rather than hours.

Determining the Rate of Cooling

The estimation of survival time based on laboratory experimentation requires extrapolation beyond data that ethical human experimentation can produce. To do this, some authors have simply extended the rate of cooling established in the laboratory down to an assumed lethal temperature. Nazi experiments at Dachau with poorly insulated individuals in stirred, very cold water provide some evidence to support this approach. In such situations heat losses are so great that metabolic heat production cannot offset them.

In circumstances where heat losses can be balanced by heat production (shivering), however, the assumption of a linear rate of fall of deep body temperature to a lethal level is erroneous. Such circumstances can occur in survivors wearing effective immersion suits, in those in life rafts, and even in some lightly clothed individuals in moderately cool water (about 20 degrees Celsius; 68 degrees Fahrenheit) provided they have sufficient subcutaneous fat. Available experimental evidence suggests that in such circumstances many individuals will, after an initial fall in deep body temperature, generate enough heat through shivering to enter thermal balance, at least for some period. In this case a simple linear extrapolation is unlikely to provide an accurate estimation of survival time (figure 7.4). Linear extrapolation based on the initial fall in deep body temperature, before thermal balance is achieved, will grossly underestimate survival time. In contrast, linear extrapolation applied after thermal balance has been achieved will predict an infinite survival time. In reality, in the latter situation the duration for which shivering can match heat lost (shivering endurance) will determine survival time.

Given that deep body temperature may stabilize at a temperature lower than that which may be achieved through ethical experimentation, it is impossible to have complete confidence in linear extrapolations from laboratory data.

Mathematical Models

Mathematical models were originally designed to simulate the human thermoregulatory system and thus increase understanding of its mechanisms and their interaction. Some of these models have been amended to estimate and prescribe the insulation requirement for immersion-protective clothing to be used in cold water. Finally, some models have been used or misused to predict how long someone will survive after immersion in cold water. These are the models used by many search and rescue organizations.

These prescriptive and predictive models are largely based on the results obtained from relatively innocuous laboratory-based experiments, in which subjects were cooled by 1 or 2 degrees Celsius (3 or 4 degrees Fahrenheit). The models perform reasonably well when validated against results obtained in such conditions. To make them fit experimental data, however, models must often incorporate correction factors, a necessity that highlights the difficulty of modeling something as complex as the human thermoregulatory system. Given the level of understanding of this system and its interactions with the macro- and microenvironment, we have no reason to believe that a definitive mathematical model of thermoregulation can be produced at this time.

Several researchers have conducted independent evaluations and reviews of models. They have used either a theoretical approach or an approach that compares mathematically predicted responses with those obtained during experimentation. The general conclusion has been that the prediction of responses in cold environments is poor, that models

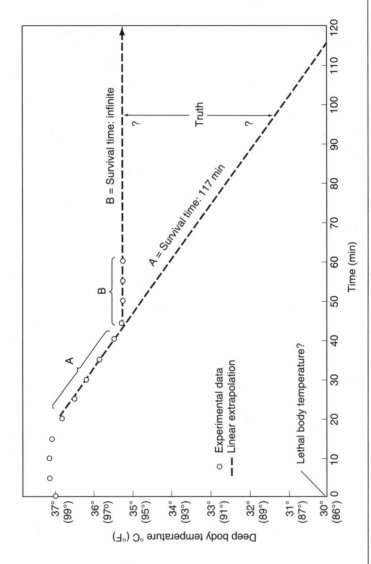

Figure 7.4 Diagrammatic representation of the error associated with employing linear extrapolation.

are suitable only for simulating a limited range of conditions, and that extrapolation beyond the range of knowledge is invalid. To sum up, models are useful as investigative tools, not predictive tools. In any case, even if a model could accurately predict the effect on survival time of variables such as sea state, gender, and age, in a real accident knowledge about the survivor may be limited to gender.

Finally, model predictions are based almost exclusively on the response of lightly clothed, young, Caucasian males undertaking immersion in relatively calm, cold water. Perhaps the greatest problem facing those who try to estimate survival time in cold water, with or without complex models, is the variability that results from differences in cooling between situations and individuals. We review briefly here some of the factors that cause this variation.

Sources of Variation

The sources of variation in the survival time of different individuals, and of a single individual on different occasions, are not just simply of academic interest. They help explain times that can vary from minutes to hours.

Sea State, Clothing, and Leakage

Conditions in open water are usually much more severe than those encountered under controlled experimental conditions. Two consequences of this are an increase in the risk of drowning and, for those wearing immersion dry suits, an increase in water leakage.

As mentioned earlier, the single most important omission from the predictions of survival is the variability caused by sea state. We discussed in the previous chapter the danger of drowning because of wave splash.

Wetting significantly reduces the insulation provided by clothing. Uninsulated dry suits keep the body warm by keeping the normal clothing worn beneath them dry. This insulation is reduced if the clothing becomes wet by leakage (Hall and Polte 1956) or cold-induced urination. Relatively innocuous laboratory-based immersion experiments have frequently generated average water leakages of $\frac{1}{2}$ to 1 liter ($\frac{1}{2}$ to 1 quart). Wetting by this volume will reduce clothing insulation by 30 to 40 percent. This equates to a reduction from 0.33 clo to 0.16 clo in the external immersed insulation provided by an uninsulated suit. Mathematical models can estimate the effect of such a reduction on deep body cooling and survival time. With the caveats noted previously and following, models predict that this decrease in external insulation will reduce survival time in water at 10 degrees Celsius (50 degrees Fahrenheit) from just over three hours to just under two hours (Hayes and Cohen 1987).

Some international specifications for survival suits have therefore been based on restricting water ingress to 200 milliliters (about 7 ounces) during leak testing.

Large volumes of water leakage will also decrease the buoyancy of a clothing assembly and increase the threat to the airway by reducing the mouth-to-water distance.

Survival curves make no allowance for where leakage and wetting occur on the body. But this can make a large difference to the consequent change in deep body temperature. A leak of 500 milliliters (one pint) over the limbs has little effect on the rate of fall of deep body temperature, whereas a corresponding leak over the torso significantly increases the rate. As cooling occurs, vasoconstriction maximizes internal insulation. Thus, in an immersed individual most of the heat loss from the body is from the torso via conduction, especially from the back of the torso because in a horizontal floating attitude the hydrostatic pressure reduces clothing thickness and, therefore, insulation in that area. This regional effect of water leakage is not observed when clothing is tested on thermal manikins rather than humans because manikins cannot vasoconstrict. Similarly, survival curves that incorporate an effect for leakage do not recognize localized effects of water leakage because manikin data have been used to determine the influence of leakage.

Largely because of increased water leakage and convective heat loss, the insulation provided by immersion suits worn in open, turbulent water can fall by as much as 33 to 100 percent from laboratory measurements. Therefore, the time of useful consciousness may be much shorter than anticipated in such situations, even in those wearing specialized garments (figure 7.5). This, in part, explains the "surprisingly poor performance of immersion suits" reported in some fatal accident inquiries.

Wearing waterproof or hydrophobic undergarments reduces the impact of wetting because hydrophobic material retains more of its thermal insulation when wet than normal underclothing. Water leakage therefore least affects immersion suits incorporating waterproof insulation in their construction. Such suits also remove the requirement for the wearer to provide his or her own insulation, but because they impair heat loss in air they can be uncomfortable to wear in warm conditions, especially when working. An insulated suit, properly worn and functioning, can reduce the rate of fall in deep body temperature by a factor of about seven during cold-water immersion when compared with normal everyday clothing (chapter 3).

Finally, even if a rescue coordinator knows that specialized protective clothing is available to a survivor, they must assume it is being worn, although even then they cannot be confident that the suit is not leaking.

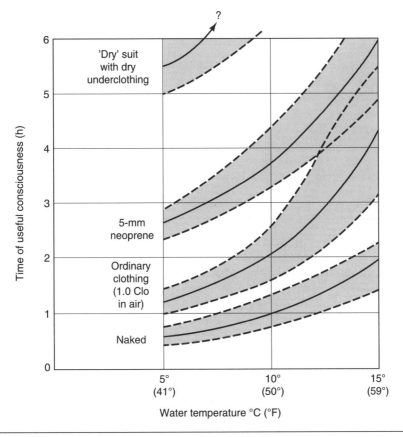

Figure 7.5   Times of useful consciousness (time to a deep body temperature of 35 degrees Celsius; 95 degrees Fahrenheit) of individuals immersed in cold water in different clothing assemblies under laboratory conditions. (1 clo = 0.155 degrees Celsius per square meter per watt, or the insulation provided by a business suit and standard undergarments.)

Buoyancy and Airway Protection

For the immersed survivor the most common sources of buoyancy are a life jacket, an immersion suit, or parts of the sinking vessel and associated flotsam. In a recent survey (UKNIIS) of immersion incidents, death occurred in 5 (3 percent) of those wearing life jackets and 45 (10 percent) of those not wearing life jackets. This latter percentage figure, however, may be much larger because the survey was directed primarily at those who survived and included only a small proportion of those who died. Life jackets save lives by helping to keep the airway clear of the water during both consciousness and unconsciousness. Unfortunately, immersion suits and life jackets have tended to be developed, specified, and tested separately. This separation has prevented the requirement for a major redesign of life jackets to be worn with immersion suits.

The question of compatibility is important as most commercially available suits will float the wearer in a horizontal attitude because of the air trapped within the garment. Although this flotation attitude is thermally beneficial—halving the surface area exposed to the water and reducing hydrostatic compression—it presents some major disadvantages in relation to the life jacket. First, the buoyancy chambers, situated over the front of the chest, remain well clear of the water and contribute little to buoyancy when the wearer is lying on his or her back in calm water. Second, the air trapped in the suit can maintain the unconscious wearer in a facedown floating position despite having inflated buoyancy chambers over the front of the chest (some life jackets have asymmetric chambers to facilitate automatic self-righting in such a situation). Third, a horizontal flotation angle tips the back of the head into the water. Many life jackets fail to alleviate this problem because their collars buckle under the weight of the head and reduce the mouth-to-water distance. Trials using a model of an unconscious human being (manikin) wearing a widely used immersion suit–life jacket combination in steep waves of just over one meter (three feet) showed that the mouth was submerged for just under one-third of the test (RGIT 1988). Immersing the back of the head has the added disadvantage of cooling a part of the body that is critical for survival (the brain stem) and readily loses heat, thereby leading to accelerated cooling of the body.

Thus, besides being thermally undesirable, a horizontal flotation angle may cause drowning. The addition of a splashguard can prove helpful in these circumstances. Often undervalued because of poor design, a splashguard can be an essential piece of protective equipment. Having one, however, does not remove the requirement for integrated, compatible designs of life jackets and immersion suits. Such an approach is essential to ensure that the head stays well clear of the water.

The problems just itemized, as well as poor fit and inferior life-jacket chamber and harness design, help to explain why some of those wearing life jackets still drown and do not achieve expected survival times.

One cannot assume, therefore, that the survival time of someone wearing an immersion suit and life jacket is a great deal longer than that of a survivor with a life jacket but without an immersion suit. The findings of fatal-accident inquiries abound with stories of people dying after one or two hours despite wearing immersion suits approved to a standard that suggested a minimum of three hours' survival in cold water. How the suit is worn, how it functions, how much and where it leaks, and how it integrates with other survival equipment, in particular the life jacket, all influence survival time. This situation will not improve until the various pieces of equipment provided for those at risk of immersion are regarded as a single integrated survival system.

The buoyancy provided by upturned hulls, large pieces of driftwood, and so on offers the opportunity for the survivor to get partially or completely out of the water. Despite the fact that it often feels colder out of the water than in it and air temperature is often lower than water temperature, in no circumstance is a survivor better off immersed in cold water. On average the cooling rate on a raft, even on one in which the survivor is exposed to the wind, is less than half of that observed in the water (Steinman et al. 1987).

Christy Ann Morrison spent 18 hours adrift after the steamer on which she had been traveling sank. She was the sole female survivor.

Photograph courtesy of National Archives of Canada.

Gender and Size

Because fat is a good insulator, when all other things are equal the average female, who has about 10 percent more body fat, should cool more slowly than the average male. But if a male and a female with the same amount of body fat are immersed, the female will generally cool more quickly than the male because females have a higher ratio of surface area to mass than males and a smaller shivering response. Thus females have a smaller heat-producing mass and a relatively larger surface area over which to lose heat. Children of either gender cool much more quickly than adults because they have a large ratio of surface area to body mass as well as less body fat.

Exercise

The effect that exercise can have on deep body temperature and consequent survival time depends on several factors, including the

▸ intensity of the exercise performed,

▸ type of exercise performed,

▸ water temperature,

▸ amount of water movement,

▸ physical characteristics of the survivor, and

▸ clothing worn.

In general, in water cooler than 25 degrees Celsius (77 degrees Fahrenheit) whole-body exercise will accelerate the rate of fall of deep body temperature compared with remaining static. This occurs because movement associated with exercise stirs the water around the body, disturbing the boundary layer and increasing convective heat loss. This effect is less pronounced if the water is already moving around the body rather than still.

Exercise also increases blood flow to the limbs. At rest, skin, fat, and unperfused muscle (resting muscle has minimal blood flow) provide insulation for the body. When unperfused, the large amount of muscle in the body can provide 70 percent of total body insulation of a resting person in cold water ("variable insulation"). At an exercise intensity of about 150–200 watts (about twice the resting level), the blood flow to the working muscles, normally those in the limbs, has increased to a point where the insulation that the muscle provides is lost, leaving only the fixed insulation of skin and fat. When muscle blood flow increases, the amount of heat delivered to the limbs also increases, much of which is lost to the surrounding water. The high surface-area-to-mass ratio of the limbs makes them ideally suited to transfer heat.

In contrast to whole-body exercise, leg-only exercise can keep deep body temperature higher during cold-water immersion than remaining at rest. This suggests that the arms are a major area of heat loss in cold water. This seems logical given that the arms have a larger surface-area-to-mass ratio than the legs, a smaller conductive pathway from the center of the limb to the surface, and smaller heat-producing muscle mass. In addition, when they are not used for exercise, the arms can oppose the torso and help insulate it.

Increases in internal insulation (fat) and external insulation (clothes) can reduce and, in the case of specialized clothing, even reverse the detrimental effect of exercise in cold water. A dry suit that does not leak essentially returns the wearer to an air environment.

We saw in chapter 4 that cold water severely impairs swimming ability. We therefore recommend that a survivor who must exercise in water use leg-only exercise and keep the arms still and close to the torso. In practice, the survivor may have to use the arms to keep the back to the oncoming waves (chapter 6).

Posture in the Water

It follows from the preceding discussion that a good posture to adopt in cold water is one that minimizes both movement and the surface area of the body exposed to the water. The natural position that people adopt as they become cold and their muscles become more spastic is a flexed fetal position. Hayward et al. (1975) have recommended that survivors consciously adopt that posture to conserve body heat while awaiting rescue. They have termed the posture "HELP" (Heat Escape Lessening Posture, figure 7.6). Stability problems make this posture difficult and impractical to maintain in an open seaway. Furthermore, the recommendation was based on the assumption that both the groin and axilla (armpits) are areas of high heat loss. This supposition has subsequently been shown to be exaggerated.

Hayward et al. (1975) have also recommended that groups of survivors should huddle together in a circle in the water (the "huddle" position, figure 7.6). They claim that this configuration will reduce the rate of heat loss, improve morale, and make it easier for rescuers to locate the survivors. Again, this advice may be suitable for the experimental tank, calm inland waterways, or lakes, but in the open sea, life will become very difficult for those on the downwind side of the group because they will be facing the oncoming waves and have to contend with constant wave splash (see the anecdote on sea canoeists in chapter 6 on pages 109-110.).

Shivering Response

Like other regulatory systems of the body, the thermoregulatory system does not act in isolation. Many other systems interconnect with it and

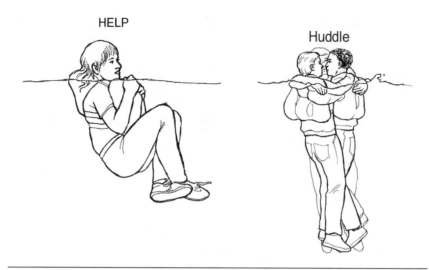

Figure 7.6   "HELP" and "Huddle" postures (not practical or recommended for open sea). Reprinted from Hayward et al. 1975.

influence it. These influences explain some of the large variation observed between the shivering responses of different individuals. Indeed, some authors have categorized subjects as shiverers and nonshiverers based on their metabolic response to the same cold stimulus. The underlying causes of this variability include inherent differences in the sensitivity of the metabolic response, age, gender, morphology, fitness, illness and injury, nutritional state, concentration of blood alcohol or other drugs, blood-sugar concentration, ambient carbon dioxide levels, ambient oxygen levels, previous exposure to cold (cold habituation), and environmental pressure (table 2.2 on page 38). These factors can influence both the threshold for the initiation of the metabolic response and its intensity. For example, age, decreased blood-glucose concentration *(hypoglycemia),* and cold habituation all delay the onset of shivering and may reduce the sensitivity of the response. These changes usually result in a faster rate of fall of deep body temperature in the cold.

Although it is relatively easy to calculate heat loss in certain well-defined situations, the numerous sources of variation make it extremely difficult to predict the intensity of the shivering response for a given individual and its consequence for deep body temperature and survival time.

Shivering, like all metabolic activity, consumes energy and can continue only while the substrates required to fuel it are available. When these are exhausted, shivering will diminish and the rate of body cooling will usually increase. Little information is available about how long people can maintain shivering at a given intensity. Shivering endurance is important because it will determine survival time in situations where a survivor can maintain thermal balance by shivering. Even the most advanced mathematical models are poor at estimating shivering endurance. Models have predicted the cessation of shivering due to depletion of muscle sugar content after four hours in water at 10 degrees Celsius (50 degrees Fahrenheit). In contrast, the evidence from human studies is that submaximal shivering can continue for at least 16 hours without food and for several days if survival rations are available (400 kilocalories plus 568 milliliters, or 19 ounces, of water per day).

Seasickness

Seasickness increases both dehydration and the rate of heat loss in a cool situation. Greater heat loss occurs because of a reduction in the intensity of the cold vasoconstrictor response and increased evaporative heat loss caused by generalized sweating. Seasickness also has a significant detrimental effect on morale and the mental state of survivors (chapter 10).

Mental State

Evidence from a number of sources emphasizes the beneficial effect that a positive mental attitude can have on the will to survive and on the determination to do what is necessary to survive. A little knowledge can help individuals retain this positive mental state. For example, the knowledge that almost all dry suits leak a little in a real survival situation may prevent the depression and panic that come with the belief that you have the only leaky suit. Doing the right things in a life raft to avoid seasickness can prevent a major cause of misery in a survival situation. The person who is still seasick despite taking antimotion illness medication must strive to adopt a positive attitude and not simply surrender to despair.

PREDICTING SURVIVAL AND SEARCH AND RESCUE TIMES

Existing predictive survival curves do little other than remind us that survival time in cold water is limited. Guidelines based on the analysis of accidents, together with laboratory-based experimental evidence, show a clear correlation between water temperature, body cooling, and survival times. But it is also apparent that because of the many factors that influence it, survival time in cold water can range from seconds to days. Predicting survival times in immersion victims is not a precise science. No magic mathematical formula can determine exactly how long someone will survive or how long a rescue search should continue. Therefore, search and rescue (SAR) coordinators must make some tough decisions based on the best information available and a number of assumptions. To cover themselves, they must extend search times beyond that which they can reasonably expect anyone to survive. Occasionally they will get it wrong, and then, in the event of litigation, it will be up to the courts to make a judgment about the definition of "reasonable expectation." A rule of thumb is for search times to be at least three to six times the predicted 50 percent survival times.

Thus, in water at 5 degrees Celsius (41 degrees Fahrenheit) the 50 percent survival time for a normally clothed individual is estimated to be about 1 hour with a recommended search time of 6 hours. The corresponding times for water at 10 degrees Celsius (50 degrees Fahrenheit) are 2 hours and 12 hours. For water at 15 degrees Celsius (59 degrees Fahrenheit), the 50 percent survival time is about 6 hours with a recommended search time of 18 hours. For water at a temperature between 20 degrees Celsius (68 degrees Fahrenheit) and 30 degrees Celsius (86 degrees Fahrenheit), search times exceeding 24 hours should be considered. Searching should continue for several days in water above 30 degrees Celsius (86 degrees Fahrenheit).

Near-naked swimmers would be at the lower ranges of these times. In calm water an exceptional individual (someone who is very fat and fit) may exceed expectations. If the rescue coordinator knows that the victim is such an individual, he or she should consider extending the search time to 10 times the predicted 50 percent survival time.

For inshore accidents, survival times may be shorter because of breaking water and adverse currents. Rescuers must consider, however, the possibility that the inshore survivor managed to get ashore. Consequently, the limiting effects of cold-water cooling will no longer be the only consideration, and the search must continue until the shore adjoining the coastline, allowing for tidal drift, has also been thoroughly searched.

Offshore, it is reasonable to expect that individuals will be better equipped to survive and have access to appropriate protective clothing, life jackets, and possibly life rafts. Consequently, search times for them should be at the upper limit, at 10 times predicted 50 percent survival time, unless obviously adverse conditions prevail.

Chapter Summary and Recommendations

> The potential usefulness of accurate estimations of survival time in cold water has, since World War II, prompted many researchers to produce predictions of survival times.

> Predictions have been based on accidents and laboratory studies supplemented by mathematical manipulation and extrapolation.

> Sources of error and variation have meant that the prediction of survival time has remained more of an art than a science.

> Sources of error include

   a. the interpretation of survival curves,

   b. the assumption of a lethal body temperature,

   c. methods used to estimate rates of cooling,

   d. use of mathematical models, and

   e. the assumption that hypothermia is the cause of death.

> Sources of variation in survival times include

   a. sea state,

   b. the effectiveness of protective equipment being used,

   c. personal factors (gender, size, fitness, health, age, shivering response),

   d. posture and exercise in the water,

   e. seasickness and mental state, and

   f. water and food availability.

> Accepting the preceding limitations, 50 percent survival times for normally clothed, young, fit, and healthy individuals approximate one hour in water at 5 degrees Celsius (41 degrees Fahrenheit), two hours in water at 10 degrees Celsius (50 degrees Fahrenheit), and six hours in water at 15 degrees Celsius (59 degrees Fahrenheit).

> Search times should be at least three to six times the predicted 50 percent survival time. In exceptional circumstances (favorable weather, well-protected survivors, and so forth), search times should be extended to at least 10 times the 50 percent survival time.

8

Necessities for Sustained Survival: Water and Food

"AFTER A WEEK, THE TERRIBLE THIRST** became a bigger problem than the general discomfort and intense heat from the sun. It was no longer simply a question of a dry mouth; now our tongues were swollen and furred, while our lips were cracked. It was difficult to muster a spit and eating our hard tack (biscuit) was impossible. After a quarter of an hour of chewing we still couldn't swallow it and in the end simply blew the powder away like dust."

World War II ship survivor

WATER

The problems for the survivor do not end with the attainment of thermal balance. The body constantly loses water, and everything it does requires energy from food. Although a lack of fluid and food does not have the urgency of the need for air and thermal balance, their absence will eventually impair performance and threaten life. Of the two, fluid replenishment has by far the highest priority.

Historical Background

The thirst and suffering experienced by people adrift differs somewhat from the more frequently documented thirst encountered by desert survivors. A tempting mass of nonpotable water surrounds the shipwreck survivor as far as the eye can see. Additionally, the survivor inhales air saturated with water vapor, and salt crystals often cover the skin and surrounding objects. Furthermore, clothes are frequently wet. The survivor may even be sitting in a pool of water, causing the skin to become swollen and puffy.

By contrast, the desert survivor is in an intensely hot (by day) and arid environment, which quickly dries the skin and every exposed mucous membrane (mouth, nose, throat). The rate of loss of water from the body is significantly faster than that encountered in a marine environment. After a relatively short time in the desert, the skin will appear to shrink and shrivel. The lips almost disappear, leaving teeth and gums protruding, and the limbs become shrunken and scrawny. Mercifully, after about three days, as the body progresses inexorably through this mummified state, death occurs. In contrast, time to death averages six or seven days in a marine environment, although, for reasons discussed later, this time can vary significantly.

The Importance of Fresh Water

"The longest reported (survival) voyage without a supply of fresh water lasted 15 days. There were three men in the boat, which was badly water-logged, and one man was drowned on the fourth day. Later a second man lost his reason and fell overboard. One survivor only was rescued. No statement was made as to whether or not rain fell" (McCance et al. 1956).

McCance et al. (1956) also reported three voyages of over 32 days in which the water supply available was sufficient to allow the men to have up to 110 milliliters (4 ounces) per head per day. Of 62 men who were adrift in two boats for periods of 37 and 49 days, and who had 110 to 220 milliliters (4 to 8 ounces) per head per day, only 1 died. It would appear then, that the maritime survivor never gets to the profound state of dehydration experienced by desert survivors. We do not know whether this is because of the aqueous nature of the environment or because the sea survivor succumbs in the end to the overpowering temptation to drink seawater before reaching a severe *anhydrated* state.

Body Water Balance

The maintenance of the optimum level of water in the body, in order to ensure normal function, is yet another example of a physiological balance: in this instance, balancing water gain against loss.

Routes of Water Loss

A resting human in a thermoneutral environment (air temperature of 28 to 30 degrees Celsius [82 to 86 degrees Fahrenheit] and 50 percent relative humidity) will lose about 500 milliliters (17 ounces) per day through insensible transudation (passage) of water through the skin (insensible skin loss) (figure 8.1). If the hand is placed inside a polyethylene bag in a cold environment, the condensation of this water on the cool inner surface of the bag quickly becomes evident. A further 500 milliliters is lost each day in saturating the inspired air in the lungs as it is heated up to body temperature. In dry hot (desert) or dry cold (polar or at altitude) conditions, this fluid loss is as much as 2,400 milliliters (81 ounces), and considerably more during exercise as respiration increases to meet oxygen requirements. In excreting the waste products of metabolism, the kidney will normally excrete a minimum of about 500 milliliters of urine per day. In ordinary day-to-day living, however, most people consume more fluid than is necessary to meet physiological requirements. The kidneys excrete the excess, giving a daily urinary output considerably in excess of 500 milliliters. Thus the minimum daily body water loss for a resting person in a thermoneutral environment will be about 1,500 milliliters (51 ounces) per day. In a survival situation, urine losses will be less as will the water gained from metabolism (see "Water Balance" on page 144). Water losses in feces are relatively small unless the individual is suffering from diarrhea. In the face of dehydration, the body will maintain circulating blood volume at the expense of intracellular and interstitial fluid volumes (chapter 2).

Clearly a number of intrinsic and extrinsic factors will influence the daily loss of water from the body (figure 8.2). Anything that increases work rate will be associated with increased respiration, body heat production, and possibly sweat loss. Sweat loss in heavy exercise may be as

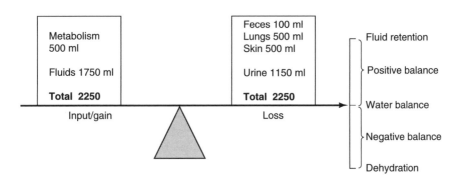

Figure 8.1   The average daily water intake and output for a 70-kilogram (154-pound) person at rest in a thermoneutral environment (water balance).

Intrinsic factors

Extrinsic factors

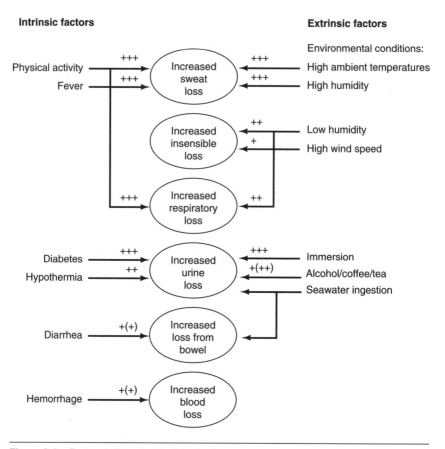

Figure 8.2 Factors influencing body water loss.

much as one to two liters (one to two quarts) per hour. People who are heat acclimatized are more likely to experience the higher rates.

Routes of Water Gain

Besides drinking, one other source of water available to the body is relevant to the survival situation—water produced within the body from the metabolism of food. This amounts to about 350 to 500 milliliters (12 to 17 ounces) per day, depending on the nature of the diet (see "Food" section starting on page 160). Both carbohydrate and fat metabolism contribute to the body water stores, whereas protein metabolism will tend to deplete those stores. A product of protein metabolism is urea, a compound that is excreted by the kidney and, in large amounts, is toxic to the body. The volume of water lost in the excretion of urea depends on the amount of protein being metabolized (2 to 3 milliliters for every gram of protein). In starvation conditions, even when a person is not eating any protein, the body begins to consume its own muscle protein *(catabolism)* as a source of vital energy.

Water Balance

In normal resting circumstances, minimum water losses will be around 1.5 liters (1.6 quarts) per day. This will be countered by metabolic water production of about $\frac{1}{2}$ liter ($\frac{1}{2}$ quart) per day. The net result is a *minimum* requirement to drink around 1 liter (1.06 quarts) per day (table 8.1). However, fluid requirements increase with normal levels of exercise and huge variations occur with additional exercise, sweating, diet, and ambient conditions.

It is difficult to set a precise value for the normal level of water in the body. Generally, it is accepted that about 65 percent of the body mass is water (70 percent of muscle mass and 50 percent of fat). Thus the body of a 75-kilogram (165-pound) person will contain nearly 50 liters (53 quarts) of water. Of this, about 2 liters (2 quarts) are regarded as surplus to requirement and act as a labile reservoir providing water to meet immediate needs without compromising essential physiological function. With the depletion of this reservoir, the brain increases production of *antidiuretic hormone* (ADH[1]) in response to the stimulation of

Table 8.1   Daily Minimum Water Requirements for a Person at Rest in a Thermoneutral Environment

Route of loss	Per day
Insensible skin water losses	500 mL
Saturation of inspired air in breathing (at rest)	500 mL
Minimal urinary losses	500 mL
Minimum daily water losses for resting person	1,500 mL
Less water gained from metabolism of average mixed diet	−500 mL
Total	1,000 mL

---

1. The authors feel that society should have an annual celebratory festival to give thanks for the antidiuretic hormone. Without it, life would be miserable and social life almost nonexistent. It is responsible for the reabsorption of around 18.5 liters (19.5 quarts) of water from the kidney per day, thus restricting daily urinary output from about 20 liters (21 quarts) per day to about 1.5 liters (1.6 quarts).

special sensors sensitive to body water concentration. In situations where a shortage of water causes the concentration of body fluid to increase, ADH acts on the kidney, lowering urine production to a minimum of about 30 milliliters (1 ounce) per hour to conserve body water. This response will occur after the loss of about 1 percent of body weight.

Minimum Daily Water Requirements

The ideal minimum requirement of one liter (one quart) of water per day and an additional quantity to balance excessive loss through sweating and respiration is unlikely to be attainable in a survival situation. Consequently, from day one of a survival voyage the body is likely to be in water deficit. The survivor can reduce the extent of this deficit by adopting a good water-conservation strategy (see page 147). The effect of the deficit will depend on its magnitude, which in turn will depend on available daily intake, activity, and weather. Therefore, it is difficult to give a blanket estimate of the absolute minimum daily requirement for a survival situation. The analysis of protracted voyages in life craft during World War II by McCance et al. (1956) found that the critical volume of potable water for survival was 110 to 220 milliliters (4 to 8 ounces) per day (table 8.2). They also analyzed the effect of voyages lasting more than three days and found broadly similar conclusions.

After the sinking of USS *Indianapolis* on 30 July 1945 in the warm waters of the mid-Pacific between the Philippines and Guam, most of the survivors had no fresh water to drink. Of the 800 men (from a crew of 1,196) who successfully abandoned ship, 484 died before being rescued, many

Table 8.2   Relationship Between the Availability of Fresh Water and Death Rate: 121 Life-Craft Voyages Involving 3,616 Men

Daily water ration per man in oz. (mL)	No. of men at risk	No. of men who died	% of men who died
"None"	143	57	40
"Some"	896	135	15
0–4 (0–110)	684	165	24
4–8 (110–220)	1,314	96	7
8–12 (220–330)	523	7	1
"Plenty"	56	1	2

From McCance et al. 1956.

from shark attack. Some of the luckier survivors managed to board open life rafts ("Carley floats") but suffered from the combined effects of exposure to the sun and dehydration. In his account of the rescue of one group of survivors, Lech (1982) describes how "they had lost about 14 percent of their body weight. . . . During the entire $4\frac{1}{2}$ days on the rafts, no one asked for a drink."

Water Conservation

In a survival situation, existing water stocks should be conserved from the outset. Survivors should consume no water in the first 24 hours because the body's reserves should be able to cope in the early stages. Much of the water drunk in this period will be excreted, wastefully, as urine. Thereafter, survivors should restrict intake to about 500 milliliters (just over a pint) a day unless supplies are plentiful, in which case they may drink up to a liter (just over a quart) a day. Overenthusiastic restriction policies from the outset may lead to the earlier onset of disability. This could reduce the ability to react to a search aircraft or other vessel responding to the Mayday or hinder effective collection of rainwater.

Survivors can reduce body water loss by optimizing the use of shade and any convective (breeze) cooling. Exposure to the sun will quadruple the loss of water through the skin. Survivors should rest during the heat of the day and, if possible, do any essential maintenance work in the cool of the evening or early morning.

In hot climates, wetting clothing and exposed skin with salt water may reduce sweat loss. The evaporation of this water will help control body temperature without the loss of precious body water as sweat. But evaporation will leave a residue of salt crystals.

The amount of protein (fish, seabirds, etc.) that it is advisable to eat depends on the amount of fresh water available. When water is plentiful, protein, which is likely to be abundantly available, provides a valuable source of energy. But when water supplies are severely limited, survivors should avoid protein as much as possible to conserve water. The loss of body water, required to excrete the urea from protein metabolism, may hasten death from dehydration long before death from starvation would occur. Thus, the benefit to body energy is outweighed by the penalty to water balance.

Given that the temperature of the sea in most regions of the world is below body temperature, bathing may help thermoregulation and reduce sweat loss. Another benefit of bathing is the opportunity it offers to stretch limbs and indulge in pleasurable gentle exercise in the relatively weightless environment of warm water. But survivors should consider several potential problems before taking a dip.

From a purely physiological viewpoint, the redistribution of fluid within the body caused by hydrostatic pressure (chapters 4 and 11) will increase urinary output, resulting in fluid loss that could exceed any gains from a reduction in sweating. By remaining reasonably horizontal in the water, however, one should be able to reduce this effect. In addition, the relatively sizable amount of energy used climbing in and out of the survival craft may cancel the potential benefit to water balance. On the practical side of the equation, getting back into the raft after spending some time in the water may well prove impossible if one is in a weakened state. The danger of suffering bites from some of the many small fish that tend to congregate beneath the shade of the craft is high. Such bites usually become infected and lead to nasty ulcerations. Attack from larger fish, including sharks, preying on these smaller ones is also possible. Finally, there is the danger of becoming separated from the craft, especially if it is drifting faster than one is capable of swimming.

Thus, the decision on whether to bathe or not will be difficult. On balance it is probably safer to adopt a conservative attitude and refrain from doing so.

Consequences of Body Water Loss

Losses in excess of about 5 percent of body weight, especially in warm environments, may be associated with headache, irritability, and feelings of light-headedness (the effect of dehydration on heat tolerance is discussed in chapter 10). The skin appears to lose its elasticity. On its release, a pinch of skin will be slow to resume its previous position. When losses reach about 8 to 10 percent of body weight, performance deteriorates significantly. The survivor may experience dizziness, faintness, rapid thready pulse, and rapid shallow breathing, possibly associated with pins and needles of the fingertips and around the mouth. Thereafter, deterioration increases, and hallucinations and delirium become common. Death usually occurs with acute losses in the range of 15 to 20 percent of body weight.

Marooned on a Yacht

In December 1987 two experienced French yachtsmen set sail from the Canaries en route to Cape Verde with the intention of crossing the Atlantic in a sturdy yacht. They unexpectedly ran into a hurricane with winds of 100 miles per hour that dismasted the yacht and partly flooded it. The salt water damaged their transmitter, preventing them from sending a Mayday. In the succeeding days, several ships passed their damaged hull but their crews failed to see the yachtsmen's distress flares.

Despite strict water rationing (they had only a 10-day supply at the outset), they ran out of water after about 50 days. The survivor recorded in his log that his companion had died of dehydration on day 55; he was so far gone himself that he could no longer remember his companion's name! By this stage he was suffering from hallucinations and had lost his fingernails and toenails. He lost 35 kilograms (77 pounds) in body weight and became delirious. On day 59, he was spotted and rescued by a German cutter. By this time he was barely alive.

Unsafe Alternative Sources of Water

In a desire to satiate an overwhelming desire to drink, many survivors turn to alternatives, some of which are dangerous.

Drinking Seawater

Accounts of the effects of drinking seawater have been compiled by Wolf (1958) and refer to such incidents as *Pandora* in 1792, *Medusa* in 1816, *Travessa* in 1923, *Rooseboom* in 1942, and *Dunara* in 1981. Lee and Lee (1971) provide additional examples from World War II. In general, these accounts suggest that drinking of seawater has an adverse effect on the body and survival.

In giving the Bradshaw lecture to the Royal College of Physicians, London, in 1942, Critchley reviewed the problem of survival at sea for sailors in the war. He stated that "seawater poisoning must be accounted, after cold, the commonest cause of death in shipwrecked sailors." Iron discipline is required to resist the ever-present temptation to drink seawater. Many eventually succumb to this temptation and drink seawater, either overtly or covertly, only to suffer the following train of events, based on composite testimony of World War II survivors who did not drink seawater:

There is an immediate slaking of the thirst, followed quite soon by an exacerbation of thirst that requires more copious draughts of seawater, and then still more. The victim then becomes silent and apathetic "with a peculiar fixed and glassy staring expression in the eyes." The condition of the lips, mouth, and tongue worsens, and a peculiarly offensive odor of the breath has been described. Within an hour or two delirium sets in, quiet at first but later violent, and if unrestrained the victim may jump overboard. If restrained, consciousness is gradually lost; the color of the face changes, and froth appears at the corner of the mouth. Death follows shortly after.

Thus, although drinking seawater may provide temporary respite from the signs and symptoms of dehydration, it will ultimately hasten death through excess sodium in the body *(hypernatremia)* before the final throes of death from absence of water *(anhydration)* occur. Maritime folklore about the danger of drinking seawater probably dates back to the days when humans first went to sea. Potable, or drinkable, water has always been a problem for seafarers, and they quickly learned the lesson of not drinking seawater, even if they did not understand the scientific basis. Invariably, the reason given was that "it turns you mad" (probably based on the hallucinations associated with hypernatremia). The historical records of survival sagas contain numerous accounts of survivors who drank seawater and, if not rescued within a few days, lost their senses and in many cases jumped overboard and died (for a historical review, see Wolf 1958).

Although ethically it is clearly impossible to obtain definitive proof that drinking seawater hastens death, the overwhelming body of scientific opinion accepts this. Nevertheless, there are some, particularly in France, who believe that drinking seawater is safe if the survivor drinks it before becoming thirsty and dehydrated. They have claimed that mixing small amounts of seawater with fresh water in the early stages of a survival voyage when fresh water is still available would help extend and thus conserve those supplies. This philosophy is based on the dubious account of an auto experiment by Alain Bombard, a French physician and yachtsman, who sailed across the Atlantic supplementing his water ration with seawater (Bombard 1953).

Before his Atlantic crossing, Bombard and a companion conducted a preliminary pilot study in the Mediterranean in a 5-by-2-meter (16½-by-6½-foot) inflatable boat, *l'Heretique,* which had a sail and rudder. They sailed from the Riviera on 25 May, and on 7 June they received food and fresh water from a ship, *Sidi Ferrugh,* at sea before landing on Minorca on 11 June. Bombard claimed that in the 14 days before picking up the provisions from *Sidi Ferrugh,* they spent 10 days without food or fresh water and the remaining 4 days living on raw fish and fish juice. He mentions that they also collected rainwater and condensation but does not state quantities. Subsequently, Bombard set out alone on the Atlantic, crossing from the Canary Islands to Barbados in 65 days. He stated that he subsisted entirely on fish juice, fish, rainwater, and seawater during this voyage.

Regrettably, it is not possible to discern from Bombard's account precisely how much seawater he consumed in proportion to fresh water. Likewise, he offers no details of how much rainwater or fish lymph he collected and consumed. Furthermore, the experiment was uncontrolled. In addition, he boarded *Arakaka* during the voyage and consumed a meal. Some time later he boarded a Dutch cargo ship, and the crew offered him some coffee! He did not state how much water, if any, he consumed, or took back on board his craft, from these two respites. On arrival at his destination, he had lost 25 kilograms (55 pounds) of body mass, was anemic, and suffered bloody *(hemorrhagic)* diarrhea, skin rashes, and lost toenails!

Although this was a remarkable achievement by a brave man, the evidence does not provide proof of his hypothesis because the exercise was not scientifically controlled. Based on the available evidence, Bombard would find it difficult to counter the charge that he would have failed without timely outside intervention. Thus the controversy continues. Let us therefore examine the underlying physiology.

When the body is depleted of water (dehydrated), the concentration of naturally occurring salts in the remaining body fluids, blood, and tissue fluids increases above normal levels. It is believed that this rise in salt levels is responsible for many of the undesirable side effects of dehydration. Excess salt in the tissue fluid bathing the cells will reduce the fluid within the cells, significantly affecting their function. It is postulated that this reduction in the intracellular fluid of brain cells causes the reported madness in those who have drunk large quantities of seawater.

An Extract From the Account of the Sinking of USS Indianapolis

"The men's thirst had become overpowering. As the day wore on, the men became increasingly exhausted and complained of their thirst. Dr. Haines (commander MC, U.S. Navy) noticed that the younger men, largely those without families, started to drink seawater first. As the day wore on, an increasing number of survivors were becoming delirious, talking incoherently, and drinking tremendous amounts of salt water. They started becoming maniacal, thrashing around in the water. These spells would continue until the man either drowned or went into a coma" (Lech 1982).

Normal body fluid has a concentration of about 0.9 percent sodium chloride (salt), which is approximately 9 grams of salt in solution per liter (1.06 quarts), a level that the body attempts to control rigidly through physiological means. Unadulterated seawater (figure 8.3), on the other hand, has a concentration of around 3.0 to 3.9 percent salt (35 grams per liter in solution).[2] Thus, if someone drinks one liter (slightly more than a

---

2. The open ocean contains 35 grams per liter of mineral matter, made up as follows: sodium chloride 77.8 percent, magnesium chloride 10.9 percent, magnesium sulfate 4.7 percent, calcium sulfate 3.6 percent, potassium sulfate 2.5 percent, magnesium bromide 0.35 percent, and calcium carbonate 0.22 percent.

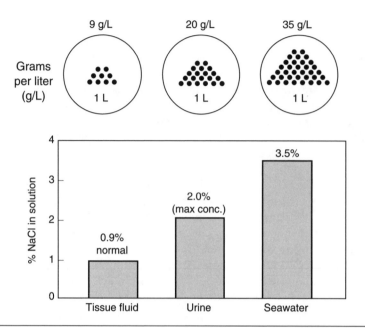

Figure 8.3 Diagrammatic representation of salt (sodium chloride) content of body fluid, urine, and seawater.

quart) of seawater, the level of salt in the body will rise above the tightly controlled level. The body will initiate a physiological response to rid itself of this excess by increasing the excretion of urine. But because the maximum concentration of salt in urine is about 2 percent (20 grams of salt per liter), after excreting a liter of fully concentrated urine an additional 15 grams of salt remains to be disposed of. This can only be achieved by excreting about another three-quarters of a liter of urine, which will further deplete body water stores. So, the result of drinking a liter of seawater is a net loss of three-quarters of a liter of body water. Alternatively, if the body cannot excrete the 15 grams of salt because it has insufficient water to waste as urine, then the salt will be retained in the body and distributed throughout the extracellular fluid (ECF; see chapter 2). This higher concentration of salt in the ECF will drag fluid out of the cells to dilute the raised concentration in the ECF. The result is that

1. dehydration becomes worse,
2. the rate of decline in performance accelerates, and
3. the onset of death approaches more closely.

It should be understood that drinking urine (2 percent salt plus urea) is also contraindicated and dangerous.

Those who promote the idea of extending limited supplies of fresh water by combining them with seawater appear to discount the normal physiological mechanism of water and salt balance. Even if the additional salt imbibed is not excreted in the urine, its accumulation in the body will increase existing salt levels, thereby causing a further decline in the water content of the cells and hastening death. Hervey and McCance (1952) examined the effects of drinking a limited quantity of either pure fresh water or a mixture of fresh water and salt water. They found that although the body retained the extra seawater, there was an accompanying rise in body salt concentration. For ethical reasons, the experiment was terminated before severe adverse effects occurred in either group.

Thus, the proponents of seawater drinking quote the incompletely documented experience of Bombard or anecdotal evidence of short-duration survival voyages (usually less than five days) in which people survived despite having consumed seawater. The opponents of seawater drinking quote the physiological facts available but are unable to provide definitive scientific evidence, which remains impossible to collect due to ethical reasons. They do, however, have some compelling supportive anecdotal evidence. McCance et al. (1956) reported on a large number of survivors in World War II, some of whom consumed seawater in their life craft while others did not. The results, shown in table 8.3, cannot be ignored, especially when it is remembered that not every occupant of the "seawater group" actually consumed seawater. Many refrained from doing so but are included in that group if anyone in their craft did so. Likewise, some of the "control" group may have consumed some seawater surreptitiously.

To counter the possible argument that the differences between the two groups might have been due to variations in time adrift (i.e., the longer the time adrift, the greater the drive to consume seawater), the authors also analyzed their data in separate subgroups according to the time adrift. The results, shown in table 8.4, remained the same. In fact, 16 of the larger control group died in days one and two, presumably as the result of injuries or drowning.

Table 8.3   Effect of Drinking Seawater on the Death Rate of Men in Life Craft

	No. of life craft	No. of men at risk	Men who died No.	%
Seawater group	29	997	387	38.8
Nonseawater (control) group	134	3,994	133	3.3

From McCance et al. 1956.

Table 8.4   Effects of Drinking Seawater on the Death Rate of Men in Life Craft: Voyages of Various Lengths

Length of voyages (in days)	Seawater group				Control group			
	Men at risk No. %		Men who died No. %		Men at risk No. %		Men who died No. %	
1–2	0	—	—	—	559	14	16	3
2–3	0	—	—	—	212	5	0	—
3–6	268	27	140	52	1,112	28	27	2
6–9	131	13	56	43	900	23	16	2
9–15	258	26	48	19	831	21	62	8
15–32	227	23	72	32	272	7	10	4
>32	113	11	71	63	92	2	2	2
All voyages	997	100	387	39	3,978	100	133	3

From McCance et al. 1956.

The lesson appears to be that to prevent a rise of salt content in the body to toxic levels, and a subsequent worsening of the state of dehydration, it is advisable to refrain from drinking seawater at all costs, either on its own or diluted with fresh water. Consuming large quantities is undoubtedly fatal, and drinking small quantities will reduce the chances of surviving a protracted voyage.

Dougal Robertson (see anecdote on page 156) makes an important point in this discussion. He comments that even if drinking a small amount of seawater was not too harmful to the body in the early stages of a survival voyage, the survivor who consumes some breaks an age-old taboo that is worth preserving. That taboo helps counter the overwhelming desire to drink seawater later when fresh water is extremely scarce.

Seawater Enemas

Some relatively contemporary survival books suggest that seawater can alleviate dehydration without ill effect if it is received in the form of an enema (colonic irrigation). The rationale is that because the large bowel is an effective reabsorber of water (preventing unnecessary wastage in the feces), seawater administered by this route will be readily absorbed, leaving the unwanted salts behind in the bowel. This technique would be valuable if there were a selective barrier to the salt being absorbed across the bowel wall along with the water. Sadly, no scientific evidence supports this claim. In fact, there is evidence to refute it. In an experiment by Foy et al. (1942), three of six male volunteers who had been deprived of water for 80 hours were given seawater enemas, while the remaining three acted as controls. The original aim was to give each of the three test subjects a 1,500-milliliter (1.5 quarts) seawater enema each day while the controls received no water by any route. In practice, the researchers found that the volunteers could not retain that volume because of violent colic and associated diarrhea (see "Osmotic Diarrhea" in chapter 10). Consequently, four daily doses of 200 milliliters (6.7 ounces) were administered instead. Even then, the subjects retained the volume only with considerable effort. Intermittent blood tests over the three days showed an increase in salt concentrations in the blood of both groups, with a significantly greater concentration in the enema group. The increase in the control group was a consequence of their dehydration and a resulting concentration of salt in the blood. Furthermore, the volume of urine production in the enema group was far higher than that in the controls, and well in excess of the volume of water administered rectally. Not surprisingly, overall body weight losses were also significantly greater in the enema group, indicating worse dehydration. The researchers concluded that seawater enemas would not be of any beneficial value and, in fact, would be positively harmful.

In another study conducted in the United States later in the same year, Bradish et al. (1942) found that the large bowel was not capable of concentrating seawater. They also noted the difficulty with retention of the enema. The volume that was retained was absorbed with its concentration of salts. These researchers also advised that seawater enemas would neither alleviate dehydration nor prolong life but instead would hasten death from dehydration.

Safer Alternative Sources of Water

Water that is potable but not palatable because of contamination from some unpleasant flavoring or low levels of salt (less than that of seawater) may be used in small doses as an enema. The Robertson family successfully used contaminated rainwater in this manner (Robertson 1973). Their story contains many valuable lessons for those who are contemplating sailing the remoter, less-crowded oceans; it is remarkable, not in the total time they spent adrift but in the hardship they endured and the innovation and ingenuity they employed to maximize their chances of survival.

The Robertsons' Tale of Survival

The Robertsons' 19-ton, 13-meter (43-foot) schooner *Lucette* was attacked by a killer whale and sunk about 322 kilometers (200 miles) west of the Galapagos Islands on 15 June 1972. The boat sank in less than four minutes, giving them barely enough time to launch their inflatable life raft and 3-meter (9½-foot) dinghy. The raft survival pack contained 10 liters (10.6 quarts) of water and food for 10 people for two days. They also salvaged 10 oranges and 2 lemons from the galley on abandonment. Their problems were compounded by having three children (age 18 and twins age 12) and a 22-year-old deckhand to provide for as well.

Wisely, they commenced a water-rationing policy from day 1, limiting themselves to one liter (a little more than a quart) per day among the six of them. On the 3rd day it rained, but the receptacles contaminated the collected water, forcing them to discard it. Thereafter, they were assiduous in cleaning the receptacles at the start of the rainstorms, so that when it rained again on the 7th day they were ready and filled all available containers. A shortage of suitable containers, however, limited their ability to collect sufficient water, so their rationing policy continued once they had drunk their fill while it was raining. By day 15, the situation was again becoming critical when Mrs. Robertson, a qualified nurse, hit on the idea of using the unpalatable water in an enema.

The raft turned out to be in very poor condition, forcing them to abandon it after 17 days of morale-sapping attention to keep it afloat. They boarded the small tender. When copious cloudbursts occurred on days 22 and 23, they devised a technique of using one of the compartments of the damaged life raft as a receptacle for surplus water. This extra water, together with turtle blood, kept them alive until a Japanese trawler rescued them on day 38.

Fish Lymph (Extracellular Fluid)

Some survival books advocate the use of fish lymph (juice, or fluid, squeezed from the flesh) to alleviate thirst. Because such fluid will have about the same salt content (concentration) as human body fluid, it will be helpful only to someone who is very dehydrated and therefore has more concentrated body fluid than normal. The energy expended and body fluid lost in undertaking the work to squeeze a small amount of fluid from fish flesh can outweigh the benefits.

Fish Eyes and Spinal Fluid

Both fish eyes and the fluid in the spinal column of fish offer another source of palatable fluid. Both the Baileys (1974) and Steve Callahan (1987) testify to the luxury of having fresh fish eyes to suck or chew when in a dehydrated state. The survivor can obtain the spinal fluid by making an incision through the spine just above the tail, with the fish held head down. By tilting the fish to the head-up position, you can collect the clear fluid in a container or suck it out. Although the volume of fluid obtained by either of these means is very little, it is satiating and has a positive psychological benefit.

Turtle Blood

Turtle blood has a salt concentration similar to that of human blood and is easy to collect—provided one is caught—and thus may help prolong survival. A turtle has about 50 milliliters of blood per kilogram (about 0.8 ounces per pound); therefore, a 20-kilogram (44-pound) animal will provide approximately one liter (one quart) of blood. In addition, stored beneath the shell of the turtle is a quantity of fat, which will provide both a valuable source of food and metabolic water. By reducing catabolism, the fat will also conserve body water.

The Elixir of Life

"I made an incision in its (turtle's) throat and collected the copious flow of blood in a plastic cup. Then raising the full cup to my lips, I tested it cautiously. It wasn't salty at all! I tilted the cup and drained it. 'Good stuff,' I shouted. I felt I had just consumed the elixir of life" (Dougal Robertson 1973).

Solar Stills

Usually inflatable, solar stills are spherical or conical in shape. An air gap separates the outer clear plastic, or similar synthetic material, from an inner dark absorbent material (figure 8.4). Solar radiation passes through the transparent material and warms the dark inner material. This inner material is wetted with seawater, which evaporates, thereby increasing the water vapor pressure in the air contained between the inner and outer materials. Some of this saturated vapor condenses on the inner surface of the clear plastic material and runs down in droplets to be collected in a channel, which feeds into a collecting chamber. This can be detached and drained periodically.

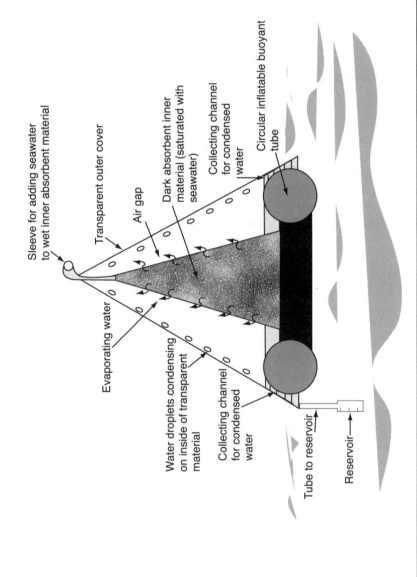

Sleeve for adding seawater
to wet inner absorbent material

Transparent outer cover

Air gap

Dark absorbent inner
material (saturated with
seawater)

Collecting channel
for condensed
water

Circular inflatable buoyant
tube

Evaporating water

Water droplets condensing
on inside of transparent
material

Collecting channel
for condensed
water

Tube to reservoir

Reservoir

Figure 8.4 A conical type of solar still.

The theoretical concept of the solar still is excellent, but regrettably, in the authors' experience, their practical performance at sea is extremely poor. The movement of the stills in a seaway makes it extremely difficult to prevent saltwater contamination of the collected moisture in some makes. But in his epic 76-day survival voyage in the mid-Atlantic, Steve Callahan found his solar still essential to survival.

Reverse Osmosis

Probably the biggest breakthrough in solving the problem of shortage of drinking water for survivors at sea was the invention and development of a device that can produce potable water from seawater through a process known as *reverse osmosis*. The principle involves pumping seawater under pressure (800 pounds per square inch) through a membrane impermeable to salt particles. Although some earlier models were mechanically unreliable and required a great deal of physical effort to produce significant volumes of water, some of the more modern types appear to be quite effective. The fluid obtained from these pumps, however, should be regarded as a supplement, rather than a replacement, for a water ration. Perhaps the best illustration of the potential benefit of such devices comes from the following account of a remarkable survival story.

The Benefit of Planning

The husband and wife team William and Simone Butler set sail from Miami on 14 April 1989 with the intention of crossing the Pacific. Two months later a school of whales rudely interrupted their dream by causing the sinking of their boat, forcing them to take to their life raft. The Butlers had fortuitously planned for such a possibility and had in their survival stores food for 30 days, 38 liters (10 gallons) of drinking water, and a reverse-osmosis pump (a Survivor 35, Recovery Engineering, Minneapolis, Minnesota). They were adrift for 66 days and owe their lives to the plentiful supply of potable water from the pump that supplemented their freshwater stores.

Rainwater

Survivors should devise a strategy to collect and retain rainwater and condensation, remembering the principle "rinse, collect, and store." Preparation should commence on day one, before weakness from dehydration occurs. Salt crystals will cover all exposed surfaces, so if they cannot be protected it will be necessary to wash them down with seawater as the rain clouds approach. Some salt will contaminate the first container filled, but it should be retained because the salt content is likely to be less than that of seawater. If the rain is protracted and all the remaining containers are full, then the first one, containing the brackish water, can be emptied and refilled with fresh, uncontaminated rain. A roll of polyethylene bags, which survivors can unfurl to collect rain over a large surface area, can prove very useful and will not take up too much space when rolled up. Having filled all available containers and satiated their thirst, survivors should use the opportunity to shower and wash all traces of encrusted salt from the skin and clothes.

Condensation

One of the sponges in the survival equipment container should be kept in a small polyethylene bag, free from contamination from salt water. This sponge can be used for collecting condensation from inside the raft. Water so collected, although not very palatable, may prove lifesaving in times of deprivation.

Additional Reserves

In remoter areas of the world, where rescue may be delayed, it is important to be prepared for a protracted survival voyage. To this end, extra reserves of water should be stowed in a readily accessible place to facilitate collection when abandoning in a hurry. Ideally, this water should be kept in watertight containers with sufficient trapped air to facilitate flotation should it be necessary to throw them overboard. The water in the containers should be chlorinated (0.2 milligrams per liter) and preferably stored in a dark place or otherwise protected from light (e.g., dark containers) to prevent plant growth. Alternatively, at routine water replenishments, they should be emptied and refilled. Overfilled containers may be difficult to locate and recover if lost overboard in a hurried evacuation.

FOOD

In 1946 the Talbot Committee (see preface), based on reports given by wartime survivors, concluded, "provided enough drinking water was available, food was of secondary importance to survivors at sea except in cold regions." Certainly, in comparison with dehydration, starvation is a less immediate threat in a survival-at-sea scenario, but nonetheless it represents a threat. Its consequences range from impairment of physical and psychological function to death. Furthermore, the availability of food has consequences for fluid balance, just as the availability of fluid has consequences for diet.

The Unfortunate Essex

In November 1820 the 27-meter (87-foot), 238-ton whale ship *Essex* was sunk by a frenzied attack from a 26-meter (85-foot) bull sperm whale in the mid-Pacific. Twenty of the crew managed to abandon ship into three small, open boats, taking with them what water and food rations they could salvage before their parent vessel sank.

As the whaling ship *Dauphin* zigzagged up the western coast of South America in search of sperm whales, the lookout spotted an 8-meter (25-foot) sailing boat in the distance. The sails, well bleached and stiffened with salt, had obviously been at sea for considerable time. As *Dauphin* drew near, those on board saw a memorable scene. The thwarts and floorboards of the small boat were littered with bones, and the two occupants sat at opposite ends of the craft, each clutching a bone from which they were sucking the marrow. The date was 3 February 1821. Instead of greeting their rescuers with smiles of relief, the survivors clutched the splintered bones with feral intensity, like starving dogs in a pit. The bones were those of some of their fellow survivors (Philbrick 2000).

Of the 20 men who abandoned *Essex,* only 8 were to survive. The 2 rescued by *Dauphin* had sailed almost 7,242 kilometers (4,500 miles), about 805 kilometers (500 miles) farther than Captain William Bligh of the Royal Navy in his epic voyage to East Timor after the mutiny on the ship *Bounty.* The story of *Essex* was subsequently to inspire the famous novel *Moby Dick.*

Energy Balance

As with many physiological functions examined in this book, starvation is also related to a balance, in this instance an energy balance (see figure 8.5). To remain in balance, and therefore to avoid starvation (negative energy balance) or obesity (positive energy balance), energy gain must match energy lost from the body. Despite both intuitive belief and apparent evidence to the contrary, the body does a good job of remaining in energy balance in the long term, in spite of considerable daily variation in food intake. As a result, the energy content of the body stays surprisingly constant. This is remarkable considering that the average person eats about a ton of food a year. If all the energy in this food were stored as fat, body weight would increase at the rate of 97 kilograms (214 pounds) a year! Instead, the average increase in body weight (fat) between the ages of 30 and 40 is about 6 kilograms (13 pounds). This weight gain is equivalent to eating three potato chips or taking two sips of beer more each day than is needed to remain in energy balance.

That energy is neither created nor destroyed during any physical or chemical process is one of the fundamental laws of science, the first law of thermodynamics. More generally known as the law of the conservation of energy, this principle applies to all living and nonliving systems.

Energy from the sun reaches the earth as radiant energy. The green pigment of plants absorbs this energy and uses it to power the production of glucose, a simple *carbohydrate* (sugar), from water and carbon dioxide through a process called *photosynthesis*. Thus, plants conserve the radiant energy of the sun. Animals feeding on plants or other animals are able to recover this energy for their own use. Therefore, the conversion of energy from one form to another occurs readily, and through that process we can trace the energy in our bodies back to the sun.

Gaining Energy: The Normal Diet

The body gains energy from the diet. An optimal diet provides the nutrients required for tissue development, maintenance, growth, and repair, as well as an adequate amount of energy for metabolic activity. The dietary sources of metabolic energy are carbohydrates, fats, and protein. For some, alcohol may contribute! The usable energy content of food made available by its total combustion is its *caloric value*.

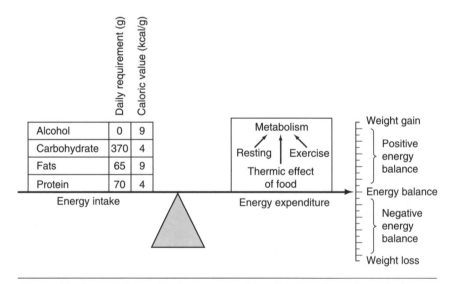

Figure 8.5 Energy balance showing causes and results of positive and negative energy balance for an average 70-kilogram (154-pound) person performing light work.

What Is a Calorie?

A calorie is the amount of heat (energy) required to raise the temperature of 1 gram of water by 1 degree Celsius. Because of the large amounts of energy involved in dietary items (a pint of beer contains about 300,000 calories!), we use kilocalories (kcal, sometimes written as Calories with a capital C). A kcal is 1,000 calories; this is more manageable and less worrisome, because a pint of beer now only contains 300 kcal! It is also the unit people will be used to seeing on diet sheets and food packaging. The International Standard unit of energy is the Joule (J); again, this is a small amount of energy so we use kilojoules (1 kJ = 1,000 joules). The relationship between kJ and kcal is: 1 kcal = 4.186 kJ, thus 1 kJ = 0.239 kcal.

The ingestion of food increases the metabolic rate of the body. The processing of nutrients by the liver causes most of this increase, and the greatest increase occurs with the consumption of protein. The average daily caloric intake of the normal diet is around 2,400 kilocalories, with 25 percent of that energy coming from fat, 12 percent from protein, and 63 percent from carbohydrate. These proportions need not be strictly adhered to because once released, the energy from the food molecules can be used to make protein, fat, or carbohydrate.

The importance of dietary fat lies in its provision of three *essential fatty acids* (linoleic, linolenic, and arachidonic acid). These are organic acids essential for specific functions in the body, including the production of lipids (see page 164). The body can synthesize some fatty acids, but the three essential ones must be eaten. Fat is also an important source of the fat-soluble vitamins A, D, E, and K. In many foods, fat carries the flavor. Fat also lubricates food, making chewing and swallowing easier. This latter point can be important to the dehydrated survivor in a life raft, for whom the monotony of the ration and the difficulty in swallowing, coupled with attenuated appetite after a few days, can make eating more of a chore than a pleasure.

Proteins are found in the cells of all animals and plants. They differ chemically from lipids and carbohydrates because they contain nitrogen in addition to other elements such as sulfur, phosphorous, and iron. Proteins are formed from subunits called *amino acids*. The body requires 20 amino acids, 8 of which it cannot synthesize. The diet must supply these *essential amino acids*; this can be achieved with a protein intake of about 0.5 grams per kilogram (0.23 grams per pound) of body weight. But the actual functional minimum, assuming normal levels of activity, is about twice that amount and is used to replace that lost by the body in hair and skin or used in enzymes (chemical catalysts) and tissue protein turnover.

Finally, a normal balanced diet provides sufficient quantities of a number of mineral substances (e.g., calcium, iron, and iodine), vitamins, and trace elements essential for the body. Vitamins are organic substances that play a vital role in metabolism, but the body cannot synthesize them in sufficient amounts. Although required only in minute amounts, a deficiency can lead to characteristic disorders, the most famous in seafaring annals being scurvy, which results from a lack of vitamin C.

For the survivor in a life raft, the normal diet just described is not something to worry about. Although shortages of vitamins, mineral substances, and trace elements will affect the body, the effect occurs after months rather than weeks, particularly in someone who starts a voyage well nourished.

Storing Energy

The whole body is an energy store because of the chemical energy locked within the structure of molecules from which it is made. Table 8.5 shows the extent of the functional store.

In contrast to plants, which store almost all of their fuel as carbohydrate (sugar), animals have evolved the ability to store fuel as lipid (from the Greek word *lipos,* meaning "fat"). The largest energy store in the body is in the form of lipid. It is an ideal cellular fuel because it is easily transported and stored and is readily converted into energy.

As can be seen from table 8.5, lipid contains about twice the energy per unit of weight as protein or carbohydrate. The ability to store fuel as lipid was a significant evolutionary advance for the efficient use and storage of energy. If, like plants, we stored most of our energy as carbohydrate rather than fat, the body weight of an average person would have to increase by about 30 percent to store the same quantity of energy. We would then have to use more energy to move about.

Table 8.5   Energy Content of an Average 70 kg (154 lb) Individual

	Total body content (kg)	Energy per gram (Kcal/g)	Total body energy content (Kcal)	%
Lipid	14.0	9	126,000	78
Proteins	9.5	4	38,000	21
Carbohydrates	0.5	4	2,000	1

About 90 percent of the lipid in the body is normally stored in fat cells just beneath the skin. Lipid is synthesized by the union of *glycerol* and three *fatty-acid* molecules. This process produces three molecules of water. Lipid is thus a concentrated energy source that is relatively water free, a characteristic that helps explain why it is such an effective insulator just beneath the skin.

Glycogen is the storage carbohydrate peculiar to mammalian muscle and liver. In well-nourished humans, approximately 350 to 475 grams of glycogen are stored in the muscles, 100 grams in the liver, and 5 grams in the extracellular fluid and blood. Each gram of glycogen contains 4 kilocalories of energy; thus the average person has between 1,500 and 2,000 kilocalories as carbohydrate. Because comparatively little glycogen is stored in the body, its quantity can be modified considerably through diet and exercise. A 24-hour fast or a low-carbohydrate diet results in a large reduction in glycogen reserves. Conversely, the dietary intake of carbohydrate significantly influences the size of the carbohydrate stores of the body and, thereby, the time to carbohydrate depletion and low blood sugar *(hypoglycemia)*.

The functional significance of these stores is that, in comparison with the carbohydrate energy store, the lipid store is just about inexhaustible. For example, a 70-kilogram (154-pound) man rowing leisurely, using about 7 kilocalories of energy per minute, would exhaust his carbohydrate stores in about $4\frac{3}{4}$ hours. In contrast, the energy in his body fat stores would be sufficient to keep him rowing for about 300 hours ($12\frac{1}{2}$ days).

As we shall see in the next section, however, the availability of the large energy store in fat does not necessarily mean it can be used.

Expending Energy

The major cause of variation in energy expenditure is exercise, including shivering. Exercise can increase energy expenditure up to 25 times the rate used at rest, shivering up to about 5 times that used at rest. Obvious physical activity accounts for only 15 to 33 percent of a person's total daily energy expenditure. The remainder sustains the body's vital functions. In contrast to the variable energy expenditure caused by physical activity, the amount of energy used to maintain the body is comparatively constant. Called the *basal metabolic rate,* it can be measured when the body is at complete rest and amounts to about 1 kilocalorie per minute (1,440 kilocalories per day). The basal metabolism is used in a wide variety of functions, including keeping the heart beating, the lungs breathing, the kidneys filtering, the liver synthesizing and metabolizing, the nerves firing, and the body repaired and in homeostatic balance.

The average energy expenditure for a human is about 2,400 kilocalories a day with normal activity. Normally, carbohydrates and fats satisfy most of the energy requirements, with the relative proportion varying depending on the tissue in question and the intensity and duration of the activity. For example, the brain normally uses only sugar. The muscles use less sugar and more fat as the duration of exercise increases, and more sugar and less fat as the intensity of exercise increases. Shivering is a form of exercise, but the hormonal *(catecholamine)* responses associated with cold stress mean that shivering uses relatively less sugar and more fat than other exercise of the same intensity.

The main function of carbohydrate is to serve as an energy fuel, particularly during exercise. On average, carbohydrates contribute about 60 percent of total energy expenditure; this can fall to about 10 percent before metabolic disturbances appear. The body must have sugar available to keep metabolizing fat and exercising or shivering, hence the saying "fat burns in a carbohydrate flame." Thus, although the energy reserve stored in fat is vast, carbohydrate must be available to use it. When carbohydrate is depleted, activity is impaired. This is one of the reasons why corpulent runners can have difficulty finishing endurance events. Fatigue occurs when muscle and liver glycogen stores become severely depleted, probably because of falling blood glucose and reduced ability to metabolize fat. As noted in chapter 2, a fall in blood-sugar levels inhibits shivering.

The oxidation of hydrogen atoms within carbohydrates, fats, and proteins provides energy for the body. Thus, the greater energy content of lipid is due to its greater quantity of hydrogen atoms. When carbohydrates or fats are broken down within the body *(catabolized)* with the consumption of oxygen to provide energy, the carbon, hydrogen, and oxygen atoms they contain end up in carbon dioxide and water. The carbon dioxide is excreted by the lungs, and the water is excreted in urine.

In contrast, proteins, particularly those in muscle, break down into amino acids (mainly *alanine* and *glutamine*). The catabolism of these amino acids produces carbon dioxide, water, and the nitrogen-containing product ammonia. Ammonia can be highly toxic to cells if it accumulates. The liver prevents this by combining ammonia with carbon dioxide to form urea, which in normal amounts is not toxic to cells. The elimination of hydrogen in this manner represents a loss of about 19 percent of the potential energy of the protein molecule. The urea leaves the liver, and the kidneys excrete it in urine. This process requires water. Approximately 50 milliliters (1.7 ounces) accompanies the excretion of each gram of urea in the urine. For this reason those suffering a shortage of water should, as mentioned previously, avoid a protein-rich diet.

The water produced by catabolism is termed *metabolic water* and provides about 25 to 33 percent of the daily water requirement of a sedentary person. The complete breakdown of 100 grams of carbohydrate, protein, and lipid produces 55, 100, and 107 grams of metabolic water respectively. Additionally, because each gram of glycogen is hydrated with 2.7 grams of water, this water becomes available when glycogen is used for energy. As discussed below, this metabolic water can become an important consideration when determining the ingredients of the survival ration.

Starvation

We must make a distinction between reduced caloric intake and no caloric intake. With planning and good fortune, the survivor in a life raft will experience the former rather than the latter. If the intake of metabolic fuels is lower than energy expenditure, the body uses its reserves of fat, carbohydrate, and protein at a rate related to the magnitude of the negative energy balance. In lean people with low reserves, a relatively large loss of tissue protein occurs when food intake is inadequate. If the deficiency continues, an increasingly serious loss of tissue occurs. Thus, the physical condition of an individual on entering the life raft determines to some extent the timing of the effects of starvation.

Adaptations to Starvation

Complete starvation leads to death in 40 to 60 days. Survival time in the absence of food depends on the ability of the body to conserve protein for as long as possible while maintaining a supply of energy to vital organs such as the heart, brain, kidneys, and respiratory muscles. To achieve this, the body reduces its requirement for energy (by decreasing metabolic rate) and lessens its dependence on carbohydrate, the most critical and limited of the body's energy stores. For survival of the brain, the body must maintain blood-glucose levels. Glycogen stores in the liver are quickly mobilized and broken down to liberate glucose. The 400 kilocalories provided by this source are enough to meet the body's total caloric requirement for four to six hours. Muscle contains more glycogen than the liver but lacks the necessary enzyme to form free blood glucose from glycogen. Instead, muscle glycogen is broken down to *pyruvate* and *lactate,* which are transported to the liver and converted into glucose. The breakdown of fat produces glycerol and fatty acids. The liver can convert glycerol into glucose, but it cannot convert the fatty acids. Instead, the muscles preferentially use the fatty acids, along with *ketones*[3] from the liver, thereby sparing glucose and protein. Virtually all the organs of the body markedly reduce their use of carbohydrate and switch to fat metabolism. After about four days the nervous system begins to use other substrates besides glucose.

---

3. Ketones are the acetone-like by-products of incomplete lipid breakdown caused when the body mobilizes more lipid than it can metabolize. This process can lead to a harmful increase in the acidity of body fluids (*metabolic acidosis* or *ketosis*).

The major source of blood glucose during prolonged fasting comes from protein broken down to release amino acids. A significant percentage of the protein of the body (in muscle, for example) is not essential for resting body function and can therefore be catabolized during prolonged fasting without serious physical consequences. The price paid is a reduction in the protein stores of the body, in particular muscle protein. In extreme cases, a significant reduction in lean body mass (weight) occurs. The load on the kidneys increases as they excrete more urea, the by-product of protein breakdown. Continued protein loss can cause dehydration, decreased function, sickness, and death.

The production of glucose from pyruvate, lactate, glycerol, and amino acids is called *gluconeogenesis*. This process can produce about 180 grams of glucose (720 kilocalories) in a 24-hour fast. The kidneys are also capable of gluconeogenesis. Indeed, at the end of a long fast (several weeks), they can be contributing as much glucose to the body as the liver is. The combined effects of gluconeogenesis and the switch to fat metabolism are such that after one month of fasting, the blood-glucose concentration falls by only 25 percent in resting individuals.

The survival value of these changes is significant. The switch to fat metabolism and reduction in glucose use means that less protein is broken down to support gluconeogenesis. Additional benefit comes from a decrease of up to 30 percent in metabolic rate caused by the loss of lean body mass (protein) and decreased activation of the sympathetic nervous system. With total starvation in an otherwise unstressed adult, lean body mass losses can approach 0.5 kilograms (1.1 pounds) per day. With the changes noted earlier, this rate is cut by a factor of four over 28 days. Consequently, protein stores last longer, and the body survives longer without serious disruption.

Physical Effects of Starvation

Research suggests that decrements in physical and cognitive (conscious) performance do not begin in well-hydrated individuals until they lose 10 percent or more of body weight. Otherwise fit and healthy individuals appear to be able to maintain normal work capacity during short periods (less than 10 days) on severely restricted diets. Again, it is difficult to separate the effects of starvation from those of dehydration. As suggested previously, however, problems that are more significant can occur from much smaller reductions in body weight when the cause of the reduction is dehydration.

If underconsumption continues for sufficient duration, changes occur in the aerobic capacity of muscle and the oxygen-carrying capacity of the blood. Stamina and physical work capacity also decline. Aerobic capacity appears to be unaffected by a reduction of up to 10 percent of body weight and thereafter falls by 4 percent for every 1 percent reduction in weight. Reduction in the mass of metabolically active tissue probably causes these changes. Loss of muscle also changes muscle biochemistry and reduces strength. Studies have reported a 21 to 24 percent reduction in maximal lifting capability during eight weeks of reduced food intake, when body weight fell by 13 to 16 percent. In contrast to these decrements, grip strength appears to stay at nearly the same level.

After the body loses significant (20 percent) lean-tissue reserves, a lesser amount of injury or stress can compromise the immune system. Resistance to infection and recovery from injury may be impaired, possibly because of the reduced concentration of the amino acids associated with immune function. Hence, skin ulcers tend to occur in long life-raft voyages (chapters 9 and 10). Long-term survivors will often observe changes in nail growth at the base of their nail beds, which take many months to grow out after rescue and a return to a normal diet.

50 days adrift! The only two survivors from the 16 who originally boarded this raft after their ship was torpedoed during World War II. One of these sailors kept a diary on a piece of sail cloth.

Photograph courtesy of Imperial War Museum. Crown Copyright.

In cold climates, metabolic rate may need to increase to maintain thermal balance. In this situation, survival time will depend on shivering endurance, which in turn will depend on the rate of utilization and provision of blood sugar. Following experiments that examined glucose utilization in cold-immersed, unfed, shivering subjects, Tipton et al. (1997) calculated that the time to cessation of moderate shivering caused by a fall in blood sugar ranged between 7 and 20 hours. These figures are in broad agreement with those that suggest that liver glycogen stores can become depleted in approximately 24 hours during starvation at rest. The consequent survival times for starved individuals who must shiver moderately to maintain deep body temperature are between 10 and 24 hours. This range is due to wide variations between individuals in body insulation, intensity of the shivering response, and size of the initial carbohydrate stores. As stated previously (chapter 6), experiments have shown that when blood-glucose levels fall to about 50 percent of normal levels, shivering lessens significantly and is even absent in some people, predisposing them to hypothermia.

The Royal Navy's daily survival ration—one imperial pint (568 milliliters) of water and 400 kilocalories of glucose ("barley sugar")—was evaluated in a laboratory cold environmental trial by Windle (1998). Eleven subjects were exposed to air at 4 degrees Celsius (39 degrees Fahrenheit) for seven days. Only four completed the trial, three withdrawing voluntarily and four withdrawing for medical reasons, including low urine output, starvation diarrhea, and vomiting. The loss of intestinal function usually causes starvation diarrhea. The subjects became dehydrated and were metabolizing their body fat and protein; their daily caloric deficit was approximately 1,450 kilocalories. They were able to maintain the light shivering necessary to maintain deep body temperature but became hypoglycemic when required to perform light to moderate exercise.

Cognitive Effects of Starvation

In the absence of mental or physical stress, body weight losses of 6 percent or less over periods of 10 to 45 days produce no degradation in cognitive performance as defined by tests of intellectual behavior (e.g., memory, reasoning, decision making, vigilance, reaction time). In the classic study of Keys et al. (1950), which was designed to induce a slow, steady, and eventually severe loss of weight, little indication of changes in the cognitive performance of the group was seen. Individual subjects did report memory lapses, inability to concentrate, obsessive behaviors, apathy, lethargy, confusion, and indifference. Psychiatric deterioration occurred in 25 percent of the subjects. Analyses of behavior during famines, of men lost at sea, and of prisoners of war show similar findings and confirm that lethargy, helplessness, and hypochondria disrupt cognitive performance. But these problems do not prevent intelligent and purposeful behavior when the opportunity arises to procure food or escape the situation. Studies that combine starvation with other stresses (such as sleep deprivation and danger) suggest that cognitive performance degradation of 5 to 35 percent can occur within days.

Survival Rations

The chance of survival will increase significantly if one is able to board a life raft containing easily accessible and appropriate survival rations.

Historically, fat has been a favorite constituent of such rations because of its high energy density and flavor. To maintain blood-glucose levels, however, a high-carbohydrate diet rather than a high-fat diet is the preferred way of providing energy because carbohydrates are digested and absorbed more rapidly than fat.

With the exception of fructose, which can cause gastrointestinal discomfort, the different types of carbohydrate sources appear to have similar value in a survival ration. Short-chain glucose polymers (as found in cereal bars derived from corn starch) break down rapidly to glucose in the body and result in a rate of delivery of glucose similar to that seen when glucose alone is consumed. Additionally, glucose polymers are more quickly absorbed in some situations, are less sweet, and may be more palatable than glucose. Storage and packaging considerations, however, may dictate the selection of carbohydrate source. With regard to the maintenance of blood glucose, solid, jelly, and liquid forms of carbohydrate are equally effective. But a solid carbohydrate bar is more satisfying and easier to preserve than a sugary drink.

Protein is definitely a poor choice as a survival food. The digestion, absorption, and assimilation of a high-protein meal elevate the metabolic rate more than a carbohydrate or fat meal. The higher metabolic rate wastes energy, and the higher level of metabolic heat adds to the physiological strain on the body in hot environments. As noted earlier in this chapter (in the section "Water"), the metabolism of protein increases dehydration. Although protein (fish and sea birds) may be the most abundant source of food, the amount the survivor should eat depends on how much drinking water is available. The more water available, the more protein one can eat, although some consideration should be given to the source!

Cannibalism on the Open Water

"Cannibalism at sea has been recorded as late as 1918 when the USS *Dumaru* blew up off Guam, and a party of survivors drifted for 23 days to the Philippines. Flesh sliced from one of the dead was boiled in a tin of seawater; the resulting broth was free from salt and not unpleasant; the meat tasted like 'tough veal.'" (Critchley 1943)

The history of survival at sea includes many examples of cannibalism. (Readers particularly interested in the legal aspects of this question should see Simpson 1994.) Regarding the above anecdote of USS *Dumaru,* it is difficult to believe that boiled seawater would taste salt free, irrespective of the source of protein with which it was boiled. In any case, apart from being ethically abhorrent, it should now be obvious that human flesh is unlikely to be of much practical value unless plenty of fresh water is available.

In 1946 the Talbot Committee recommended that life-raft rations should be about 1,000 kilocalories per day, per person, for four days. The proposed 1,020-kilocalorie ration comprised 500 kilocalories of high-fat biscuit ("cookie"), 200 kilocalories of sweets (candy), 150 kilocalories of fruit block, and 170 kilocalories of sweetened condensed milk. With some minor alterations in quantities, the Royal Naval Life Saving Committee accepted the ration on an interim basis and used it to provide the size and weight of the ration in the new life rafts.

In recognition of the fact that the dietary intake of carbohydrate significantly influences the size of the carbohydrate stores of the body, the Royal Naval Personnel Research Committee recommended three levels of survival rations for life rafts in 1950. The first was the desirable ration of 1,730 kilocalories per day, the second was the compromise ration of 1,250 kilocalories per day, and the third was the minimum ration of 600 kilocalories per day for five days. The committee estimated that men who were still adrift in the Tropics or the Arctic after living on the desirable or compromise ration would all still be alive, but some with the minimum ration might be dead. Subsequently, constraints on space resulted in a reduction of the life-raft ration to 400 kilocalories (100 grams of barley sugar [glucose] sweets) and an imperial pint (568 milliliters) of water per person per day for five days. Survivors are advised to drink nothing on the first day, until the body adjusts to the consequent mild dehydration by reducing the excretion of urine.

In the study by Windle (1998) described previously, the 11 subjects who had been exposed to air at 4 degrees Celsius (39 degrees Fahrenheit) for seven days reported that they did not like the barley sugar and would have preferred a variety of foods including chocolate, cereal bars, and flapjacks. In a previous field trial in a life raft at sea, however, subjects did not find the absence of food a problem and ate barley sugar sweets without complaint, although the minimal water rations were a source of concern. Perhaps the slightly warmer environment and more uncomfortable conditions made food, or its absence, seem less of a problem in the open sea trial.

McCance et al. (1956) examined the reports of survivors from 461 voyages in boats and life rafts. Of these, 70 percent had been in the North Atlantic. In 149 instances, survivors had noted details of the type of food eaten as well as its acceptability. On 34 voyages lasting 3 to 77 days, 300 to 500 kilocalories per day had been available, and many people survived even in low ambient temperatures. Many found cookies to be too hard and dry. Some liked chocolate, but others did not. Survivors generally appreciated sweetened condensed milk, but it made them thirsty. Contrasting findings were obtained from trials conducted off the south coast of England in 1960, in which 16 men in a 20-man Royal Navy life raft received 400 kilocalories a day for five days. In this trial, cookies were the preferred food, followed by barley sugar sweets and milk tablets. The subjects least liked rum fudge and condensed milk. The fact that a pint of water a day was available to each person in the latter trial can explain some of the discrepancy between these and earlier findings.

Hervey and McCance (1952) showed that replacing 100 milliliters of a 350-milliliter water ration with 100 grams of carbohydrate (barley sugar sweets) conserved 200 milliliters of body water by the third day of the trial. The improvement occurred because of a reduction in urine volume, which they attributed to a sparing of body protein, prevention of ketosis, and a reduction in metabolic rate.

One should remember that survival rations are not simply meal substitutes. Their primary purpose is to provide sugar in order to reduce catabolism and dehydration and thereby extend survival time. Survivors should regard survival rations more as a medicine than a meal. Although they may have little appetite for the rations after several days adrift, survivors should still consume them.

Within the confines of space and weight limitations, the survival ration should comprise a variety of foods composed predominantly of carbohydrate. These might include chocolate, cookies, oatmeal block or other high-energy bars, and boiled sweets. The daily ration should provide between 600 and 1,400 kilocalories. As a guide, the caloric content of six cookies (84 grams) is about 332 kilocalories; of oatmeal block (25 grams), 124 kilocalories; of a bar of chocolate (60 grams), 315 kilocalories; of a packet of glucose sweets (50 grams), 200 kilocalories.

Finally, the survivor should remember that reducing activity to a minimum will decrease the requirement for food (energy) and water.

Chapter Summary and Recommendations

▸ Fluid and energy balance are intimately related. Their maintenance can be critical to performance, health, and survival.

▸ Dehydration in excess of about 5 percent body weight may be associated with headache, irritability, and feelings of light-headedness. With losses of 8 to 10 percent, performance declines significantly. Further losses lead to hallucinations and delirium. Death usually occurs with acute losses in the range of 15 to 20 percent of body weight. In a marine environment this occurs in 6 to 7 days

▸ In well-hydrated individuals, physical and mental capabilities do not decline until body weight loss exceeds 10 percent. Death from starvation takes 40 to 60 days.

▸ For the average resting adult, daily fluid loss is 1,500 milliliters (about $1\frac{1}{2}$ quarts), and daily energy expenditure is 1,400 kilocalories. The recommended minimum daily requirement for fluid is 1 liter (about 1 quart), and for energy it is 1,400 kilocalories. In a survival situation in optimal conditions, these quantities may be reduced to a daily intake of 110 to 220 milliliters (3.7 to 7.4 ounces) of water and 600 to 1,400 kilocalories for a limited period. To reduce catabolism and dehydration, these calories should be in the form of carbohydrate.

▸ The survivor can reduce food and water requirements by minimizing energy expenditure and water losses. This can be done by

 a. drinking nothing in the first 24 hours,

 b. never drinking seawater,

 c. never mixing seawater with fresh water,

 d. avoiding eating protein unless fresh water is freely available,

 e. minimizing activity,

 f. resting during the heat of the day and working in the cool of the evening or early morning,

 g. optimizing the use of shade and breeze, and

 h. employing "artificial" sweating (wetting) when appropriate.

▸ The potential life-raft survivor should consider alternative means of acquiring water (for example, ways of collecting rain and condensation; reverse-osmosis pumps and solar stills; fish lymph, spinal fluid, and eyes; turtle blood).

▸ Those at risk of becoming life-raft survivors should prepare emergency water containers for life craft.

Fat reserves are plentiful, but glucose is required to enable the metabolism of fat. Protein reserves are also reasonably plentiful and can be used to provide the glucose to light the "flame to burn" the fat, but muscle wasting and protein deficiency disorders quickly follow. A minimal daily intake of carbohydrate will help offset this.

The absence of vitamins, minerals, or trace elements is unlikely to pose a problem to life-raft survivors in the short term (two months).

9

Castaways: Survival in an Open Boat or Life Craft

The term "castaway" generally refers to a shipwreck survivor in some form of survival craft rather than an individual who has been marooned. Surviving the sinking or abandonment of a ship or boat is only the first step to surviving a disaster. Surviving at sea beyond that requires more than having the appropriate "survival equipment"; it is also necessary to have knowledge, training, plus tangible goods like food and water. Other factors high on the list of requirements are traits like fortitude, ingenuity, and hope. Without such a combination of factors, only the lucky will survive.

The Plight of the MV Lovat

In January 1975 in a southwest gale (force 8 to 9), with heavy, confused seas running, the cargo of coal slurry began to shift in the 581-ton MV *Lovat* when she was just 29 kilometers (18 miles) off the southwest tip of England. At 0620 the ship took on a severe list to starboard; a Mayday was transmitted and acknowledged by Lands End Coastguard. At 0630 the order to abandon was given. The 13 crew members took to the 10-man inflatable life raft because the severity of the list prevented the ship's lifeboat from being launched. The raft was of the type that had two buoyancy tubes, one superimposed on top of the other. Unfortunately, in the 15 or so minutes that it took to launch, maneuver, and board the raft, the lower buoyancy tube was

(continued)

(continued)

abraded and began to leak so that by the time all 13 had boarded the raft, the tube was no longer fully inflated. Consequently, the waves were washing in an open entrance on one side, and out the opposite entrance. Inside the raft these waves were rebounding off the inside of the canopy, creating problems with breathing for the crowded occupants, who by now were sitting waist deep in cold water (about 8 degrees Celsius; 46 degrees Fahrenheit) in ordinary clothing. Nevertheless, initially all the occupants were in good spirits with the standard ribald comments about where others were putting their feet!

None of the occupants appeared to have experience or training on how to optimize the use of their raft. The sea anchor was not streamed, and although an attempt was made to close the canopy, a combination of cold hands and unfamiliarity prevented it from being achieved. Using their shoes they tried to bail, but because the water was washing in faster than they could bail, they soon abandoned the effort. At this stage, in an effort to improve conditions inside the raft and enhance the chances that all would survive, they agreed to lighten the load by taking it in turns to enter the sea. They were confident that rescue would not be long in coming because they were only 29 kilometers (18 miles) offshore and in a busy shipping lane.

The 15,500-ton container ship *Discoverer,* alerted by the Coastguards of *Lovat*'s plight, proceeded to the position of the incident and arrived alongside the heavily listing ship at 0705. They found the starboard gunwales awash but no sign of life. *Discoverer* commenced a search for survivors. *Lovat* eventually sank at 0730, 45 minutes after the crew had abandoned.

At about 0745, in the gloom of early dawn, the raft occupants sighted *Discoverer* close by ("within shouting distance"). They scrambled to open the life-raft survival pack looking for flares. Unfortunately, by now the toughness of the polyethylene wrapping proved too much for their numbed, cold hands; none of them were aware of the knife in a scabbard on one of the canopy support arches. The ship passed by twice more without sighting them. It was later established that the raft's sea cell light was not functioning. By now many of the occupants were semiconscious from the cold. About 30 minutes later, a high-sided car ferry located them and came alongside, but those in the raft were unable to catch the line thrown to them. All the ferry could do was to keep them in its lee and act as the on-site rescue coordinator. Fifteen minutes later (0830), a small rescue helicopter arrived. At this stage the raft was partially submerged with the four remaining conscious survivors standing, two to each entrance. Two dead bodies were floating inside the partially flooded raft; the sea had washed out the others.

In the 40- to 50-knot wind, with the raft rising and falling 9 to 12 meters (30 to 40 feet) with each wave peak and trough, the helicopter, using a rescue diver, commenced a double lift on the first survivor. Unfortunately, at a critical moment just as the hoist was commencing, a particularly large wave resulted in some slack in the winch wire becoming entangled around the body and neck of the rescue diver. Before he could free it the wave began to recede, coincidental with the lurching of the helicopter to one side from a sudden gust of wind from an eddy caused by the ferry. The winch operator had no option but to cut the wire. The diver disentangled both himself and the victim from the wire and helped the survivor back to the raft to await the arrival of a bigger helicopter. By the time it arrived, about 40 minutes later, only 2 of the original 13 had survived the $2\frac{1}{2}$-hour ordeal.

After this incident the British government's Department of Trade instituted an obligatory sea-survival training program that all *ab initio* seamen must complete before being certified for a seagoing job.

HISTORICAL ANECDOTES

Without an analysis of a large number of such disasters worldwide, it is difficult to obtain a global picture of the real problems confronting survivors at sea in survival craft. Generally, the more dramatic incidents make front-page headlines and influence public opinion. Frequently, these are harrowing accounts of lone sailors struggling for weeks against the elements after their yachts have sunk, a relatively recent example being the excellent account by the lone yachtsman Steve Callahan, who spent 76 days drifting in the mid-Atlantic in a life raft (Callahan 1987). Such examples provide a good indication of the nature of the problem in a particular geographical area. But focusing on these kinds of incidents can bias the overall picture, because the balancing evidence of those who do not survive is clearly unavailable. For that, one must turn to the excellent study of Professor McCance and his colleagues after World War II (McCance et al. 1956).[1] They examined the depositions of survivors from 448 ship sinkings, mostly British merchant ships, in 1940 through 1944. These worldwide incidents involved 27,000 men. Their conclusions state:

▶ About 68 percent of the men were rescued. Some 26 percent were lost before they reached some form of life craft, but relatively few

---

1. For an excellent, nonmedical, historical overview of the problems facing British merchant-shipping survivors in World War II, read *Survivors* (Bennett and Bennett 1999).

were killed or trapped by damage to the parent ship. A further 6 percent of all those at risk died after they reached the life craft.

▸ Cold, intensified by exposure, was the single most important cause of death before and after boarding the life craft. On short voyages in sea temperatures below 5 degrees Celsius (41 degrees Fahrenheit), the death rate was 20 to 30 percent, whereas it was less than 1 percent on short voyages above 20 degrees Celsius (68 degrees Fahrenheit).

▸ High mortality occurred on long voyages. Only 2 percent of the men who reached the life craft died if they were picked up by the second day, but 26 percent died when they were adrift for more than 15 days (figure 9.1). This tendency was more marked in colder waters (table 9.1).

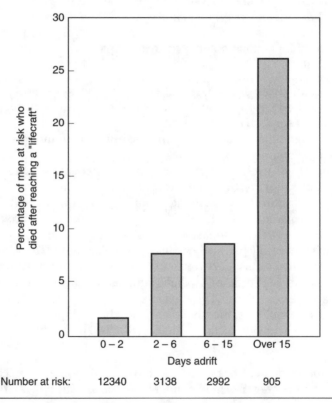

Figure 9.1 The effect of time adrift on the death rate at all water temperatures, boats and rafts together (in 461 life-craft voyages).

From McCance et al. 1956.

Table 9.1   Effect of Sea Temperature on Death Rate (%) During Survival Voyages of Up to Six Days Duration in Survival Craft (Boats and Open Rafts)

	Sea temperature °C (°F)			
	<5 (41)	5–10 (41–50)	10–20 (50–68)	20–31 (68–88)
% of men at risk who died	50	36	6	6
No. at risk	306	1,240	7,894	6,101

From McCance et al. 1956.

▶ Relatively few of the life craft at risk were never picked up or failed to make shore. Only 2 percent of all men at risk in life craft were lost without trace.

▶ More frequent and more realistic drills, better means of launching and boarding life craft, and better protection against the elements would have saved thousands of lives.

▶ On life-craft voyages longer than three days, the drinking of sea-water was accompanied by a rise in death rate from less than 4 percent to about 40 percent. The provision of sufficient fresh water would have prevented this.

▶ Better conspicuity and signaling devices would have shortened a number of the life-craft voyages and so materially reduced the loss of life.

From the foregoing it is readily apparent that the two great physiological threats to survivors in life craft are cold conditions and insufficient drinking water. In their sample, McCance et al. (1956) found cold to be by far the most serious threat, which is perhaps not surprising as more than 57 percent of the ships in their survey sank in the North Atlantic. They conclude, however, that the death rate from cold would have been even greater if these ships had not been sunk while in close contact with others in convoy or near shore. Most of those who were not picked up early succumbed to the effects of cold long before thirst became a critical problem. In fact, only one man adrift at a sea temperature below 5 degrees Celsius (41 degrees Fahrenheit) survived for a period exceeding 15 days. In the 268 voyages exceeding 15 days at a sea temperature between 10 and 20 degrees Celsius (50 and 68 degrees Fahrenheit), the fatality rate was 52 percent. The corresponding figure for the 199 voyages in sea temperatures between 20 and 31 degrees Celsius (68 and 88 degrees Fahrenheit) was 13 percent.

Because of the lessons learned from World War II, physiologists and naval architects undertook collaborative research to devise a means of reducing the threat to survivors at sea. Improved life jackets, survival suits, and life rafts were the result. Continuing research and advances in material technology have improved both life jackets and survival suits in recent years. Sadly, however, many lessons in life-raft design, acquired at such cost, have been ignored or forgotten as economic pressures and vested national interests have influenced international regulatory policy. Too often, such regulations are vague, poorly thought through, or set at the lowest common denominator. A practical consequence is the proliferation of equipment that meets the letter of such regulations but lacks functionality. Thus we see life jackets that carry the obligatory instructions for inflation in a location not visible until the life jacket is inflated. We find life rafts that include the requisite canopy, but it adds little to the thermal protection of the occupants. Helicopters may carry the required rafts, but little thought is given to ensuring that they are available after ditching.

Trouble With Life Jackets

"Many passengers reported difficulties with the life jackets. One passenger said they were tied together in threes and were difficult to separate. Others found the straps too short to be fastened at the crotch. Most witnesses did not understand how to put the life jackets on, they did not seem to fit. Some reported that the straps were missing or too short. Many witnesses lost their life jackets when they jumped or were washed into the water and several reported that the jackets slid down around their waists" (MV *Estonia*, Final Report 1997).

When replacing their old life rafts in the 1970s, the British navy conducted extensive trials on the large commercial life rafts then available. The type of raft the navy eventually chose required extensive modification before acceptance and introduction into service. Fortunately, all the ships involved in the Falklands conflict in 1982 carried these modified rafts.

Falklands Experience in 1982

Given the known destructive power of modern weapons and the materials from which modern warships are constructed, casualty predictions for war at sea are high. It was with some surprise, therefore, that in the

Falklands conflict in 1982 the number of maritime casualties was so low. On the British side, only 6 percent of personnel on board ships that were sunk while underway were killed in action, and less than 1 percent died in the abandonment-survival phase (table 9.2). This figure is in stark contrast to the 67 percent that died during this phase in World War II.

The data are a little misleading, however, because personnel in only two of the ships *(Coventry* and *Atlantic Conveyor)* were obliged to take to the water. In the others, most survivors were able to step onto another ship that had come alongside to their assistance. Even so, the loss of life among the survivors who were obliged to take to the water was impressively small given the circumstances. In both sinkings, individuals abandoning were often shocked and dazed. Many were exhausted from protracted attempts at damage control before abandonment. Some were injured and burned. All were eventually forced to abandon in a hurry into very cold water (about 5 degrees Celsius; 41 degrees Fahrenheit), with the ship either about to capsize or uncontrollably on fire and with ordinance exploding all around them. This success was largely the result of the excellent personal protective equipment provided. Every man had a personal quick-don survival suit and life jacket, and the life rafts were developed after extensive environmental evaluation and testing. Sadly, poor training and unfamiliarity with equipment was the possible cause of the deaths of some of the survivors. Had people been obliged to spend a longer time afloat in the life rafts before being rescued, the loss of life would undoubtedly have been considerably greater, although it is

Table 9.2   Number of Personnel Who Were Killed in Action or Who Died During the Survival Phase in British Ships Sunk While Underway at Sea in the Falklands Conflict, 1982

Ship	Number of crew	Killed in action	Died in survival phase
HMS *Sheffield*	300	20	—
HMS *Coventry*	300	22	2
HMS *Ardent*	202	22	—
HMS *Antelope*	175	1	—
Subtotal	977	65	2
Atlantic Conveyor	149	3	9
Total	1,126	68	11

unlikely that it would have reached the proportions reported in World War II by McCance et al. (1956).

The unfortunate Argentinian survivors of the *Belgrano*, sunk by torpedo, were less lucky. Of the 1,000 crew (approximately), none of whom had survival suits, 640 survived. An Argentinian ship, *Aria Bahia Pariso*, picked up 89 "survivors" in life rafts, of which 71 were alive and 18 dead. Of the 71 alive, 69 were suffering from hypothermia, thus confirming that in many modern life rafts, even in a relatively short time adrift, the thermal problem remains the predominant threat to life.

Surviving in Extreme Conditions: Shackleton

Given that cold is the major threat during both short- and long-term survival voyages outside the Tropics, the epic survival voyage by Sir Ernest Shackleton and his crew from *Endurance* is all the more remarkable. Of all the many harrowing accounts of successful survival at sea, few can compare with Shackleton's, and it is doubtful that any will ever surpass it. To anyone interested in survival, the account of Shackleton's voyage to safety, with its unimaginable physical and emotional demands, is essential reading. The story should serve as a model for all never to give up, regardless of how impossible a situation appears to be.

Some may argue that the Shackleton story is not representative of the problems facing genuine castaways. After all, Shackleton and his crew had time to prepare for their ordeal and ample reserves of appropriate equipment and stores. In contrast, most sea survivors are obliged to abandon ship in a hurry. Nevertheless, the position of Shackleton and his men was extremely precarious, and no one can deny that their sea voyage was a survival one. Furthermore, most of their equipment was designed for living in the cold and aridity of Antarctica and was therefore quite unsuitable for survival in a cold, wet marine environment.

By judicious planning, preparation, and above all inspirational and innovative leadership, Shackleton optimized the equipment available to him and led his men to safety despite the appalling odds against them. In far less severe conditions, many people have failed to survive even 24 hours in temperate climates, despite the availability of specially designed sea-survival equipment. If one plots a sample of the recorded survival-craft voyages on a global map, it is apparent that the long-duration voyages are located almost exclusively in tropical or subtropical waters, with Shackleton's one of the few exceptions (figure 9.2).

This figure provides dramatic evidence that problems associated with the maintenance of thermal balance in cold climates (in temperate as well as arctic environments) are one of the major limiting factors to successful survival at sea outside the Tropics. Furthermore, cold is probably a greater threat to survival at sea than lack of potable water. Failure

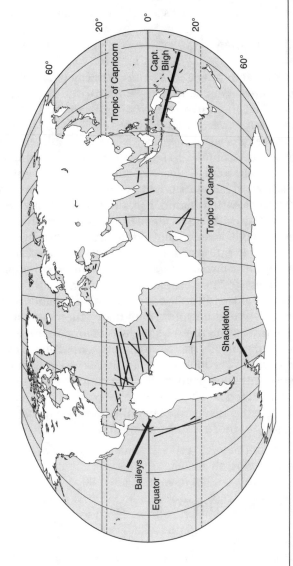

Figure 9.2 Some of the long-duration survival voyages recorded in the literature shown on a global map (adapted from the original by G.R. Hervey). Note that the majority occur in subtropical or tropical waters.

to maintain body temperature can kill in less than 24 hours in temperate climates, whereas the lack of potable water is unlikely to be fatal in less than six days, even in the Tropics.

One of the few environmental threats that Shackleton did not have to endure is the unremitting, blistering heat of the sun. This, as we see in the following section, can present a special set of challenges.

Longest Time Adrift: Poon Lim

The record to date for time spent adrift as a castaway (130 days) goes to Poon Lim, a Chinese crew member from the British cargo ship *Ben Lomond* that was torpedoed off the northwest coast of Brazil on 23 November 1942. After his ship sank quickly, he managed to board a "Carley float" from the water. These rafts were the principal survival aids for the crews of warships at that time. Some were supplied to merchant ships during the war to provide an alternative means of survival in the event the ship sunk before the standard lifeboats could be launched. These rafts consisted of an oval outer flotation chamber, made from wood, cork, or occasionally metal, painted gray. A lattice, made from netting or other material, straddled the central void. Fixed in the middle of this netting was a small container with survival rations and flares. Most of these rafts were designed for 20 persons, although some of the larger versions had a capacity of as much as 50. The occupants could be in the water holding on to a lifeline looped around the outer perimeter or standing inboard on the netting, which would be submerged under their weight. If there were only a few survivors and the sea state was favorable, they could sit on the buoyancy chamber of the raft with their legs dangling in the water. The rafts had no means of propulsion and provided no shelter from the elements; they were intended primarily as a means of temporary refuge until another ship effected a rescue. Sadly, theory rarely matched practice, and many thousands of survivors perished on these rafts from cold or dehydration. Many of those who survived suffered severe cold injury to their feet.

Dying Aboard the Carley Float

Of the 2,000 or so men aboard the aircraft carrier HMS *Glorious* when she was sunk off Narvik on 8 May 1940, about 400 managed to board Carley floats. Only 60 were rescued alive. All had cold injuries, and some died later in hospital. "Tubby" Healiss and one other were the sole survivors from a Carley float that originally had 50 survivors (Healiss 1955).

Survivors from a Carley float scrambling up a rescue net suspended from a boom extension on the bow of a dedicated convoy rescue ship in 1943.

Photograph courtesy of Imperial War Museum. Crown Copyright.

Fortunately, in Poon Lim's case, the sea was warm and, being alone on such a raft, he had ample water and food, at least for some weeks. Furthermore, the frequent tropical downpours common in that part of the ocean at that time of the year replenished his water supplies at regular intervals. After 50 days, however, his food supplies began to run out, although he managed to catch some fish to reduce the level of starvation. His biggest problem was the unrelenting sun that blistered his skin horribly. Nevertheless, he remained hopeful and managed to keep his spirits up. After 120 days the pilot of a passing aircraft indicated to him that he had been seen. Unfortunately, it was another 10 days before a fisherman rescued him, 16 kilometers (10 miles) off the coast of Belem, Brazil. By that time he was extremely emaciated and could not move to help himself. Surprisingly, after 15 days in hospital in Belem he recovered and was fit for discharge. His 130-day ordeal remains a record to this day.

Record Time Adrift in Peacetime: The Baileys

The next longest survival voyage is that of the husband-and-wife team of Maurice and Marilyn Bailey. A whale holed their yacht on 4 March 1973 about 483 kilometers (300 miles) north of the Galapagos Islands (Bailey and Bailey 1974). Their attempts to seal the hole in the hull were in vain, and eventually they were obliged to admit defeat as the boat began to sink beneath them. Before abandoning, they had time to gather a large supply of drinking water, some provisions, and other oddments that they thought might be useful. Their use of this time demonstrated remarkable insight into the problems that they were subsequently to encounter and should serve as an object lesson for many. As their yacht sank they began their voyage in an inflatable raft just 1.4 meters (4½ feet) in diameter with an inflatable dinghy in tow. Little did they know they were about to enter the record books. After 118 days adrift, the crew of a Korean fishing boat spotted and rescued them on 30 June 1973. Their story is another remarkable account of endurance and fortitude in the face of thirst, carbohydrate and vitamin deficiency, chronic saltwater ulcers, and cramped conditions, coupled with despair and disappointment as ships passed nearby without sighting them.

A Family of Survivors: The Robertsons

By an extraordinary coincidence, when the Baileys were in the middle of their survival voyage, one of the authors was interviewing Dougal and Lynn Robertson, another husband-and-wife team who had a harrowing survival experience in the same part of the Pacific a year earlier. Their survival voyage, although considerably shorter (38 days) than that of the Baileys, was possibly more stressful. They were short of water and had their young children to worry about as well as themselves (chapter 8). The story of their harrowing voyage (Robertson 1973) and subsequent survival manual (Robertson 1975) are essential reading for those contemplating sailing in the more remote seas of the world.

Dying From Cold in the Sub Tropics

Even in the relatively warm waters of the Azores, the problems of survival are not restricted to a shortage of potable water, as the following account reveals.

The Pamir

On Saturday, 21 September 1957, the German sail training ship *Pamir,* with a crew of 86, was overcome by hurricane Carrie about 805 kilometers (500 miles) southwest of the Azores. A number of survivors managed to clamber aboard lifeboats or rafts from the water. Although several rescue ships closed on the scene in response to her call for assistance, two days passed before any survivors were rescued. In those 48 hours the survivors saw many ships pass close by—one as near as 91 meters (100 yards)—but with the high prevailing sea state and the absence of suitable location aids, those in the ships did not see them. By Monday evening, a damaged, partially flooded lifeboat with 5 survivors was located just before sunset. The boat had originally contained 10. Two died from exposure in the first 24 hours and 2 others, in spite of being told not to, drank seawater and later that day "went mad, jumped overboard, and swam away." On Monday the remaining survivors became increasingly despondent because of the number of ships they could see that evidently could not locate them. They constructed a makeshift mast from which they draped material from some unused life jackets. This improvised item eventually led to their being located. On Tuesday afternoon, a lifeboat with a lone occupant was sighted, the sole survivor of 20 who had originally boarded that craft on Saturday.

In all, there were only 6 survivors of *Pamir.* Several life jackets were subsequently found with ties unfurled and knotted, suggesting that they had been used at some stage. Some inflatable life rafts, with evidence of occupation, were also located.

The remainder of this chapter is based on the lessons learned from such anecdotes, personal practical experience combined with experimental results, and an understanding of the underlying physiology described in the previous chapters.

LESSONS LEARNED FROM HISTORY

The preceding stories encompass all the major hazards confronting the survivor adrift at sea. They demonstrate opposite ends of the environmental spectrum and the differing challenges they present. Furthermore, they are inspirational and highlight the importance of fortitude and the will to survive. They also show that even when safety or rescue is within sight, there is no guarantee that it is imminent. The adage that "no one is a survivor until he or she has been rescued and made a full recovery" is worth remembering. We shall return to it in chapter 11.

The survivor must therefore maintain discipline to the end and avoid becoming demoralized if things do not immediately work out favorably. In addition, unless one has good working knowledge of the nature of the threat, of how to use the available equipment, and of how to improvise when necessary, then survival is going to be difficult, if not impossible. Survivors must remember that although they may see a rescue craft, those in the rescue craft may not see them. The accounts demonstrate that several factors are essential to survival. Those who survive have usually made specific preparations and have the following attitudes and knowledge.

▶ They never assume that a survival situation will not some day occur.

▶ They understand and respect the nature of the threat.

▶ They recognize that those who survive are not only lucky but also well prepared (that is, they make their own luck).

▶ They plan safety and survival strategies and make preparations accordingly, in advance.

▶ They review those plans regularly and revise them as necessary in the light of changing operational, seasonal, or geographical circumstances.

▶ Their emergency "grab bag" has the appropriate contents for the environment in which they are operating.

▶ They have a "reserve list" of supplementary items to take if time permits.

▶ Their equipment is regularly serviced and up to date.

▶ They practice the drills and understand how to handle the life raft in the water.

▶ They understand the physiological principles involved to be able to improvise as necessary.

▶ They remember that as long as a boat or ship remains afloat, it is the best lifeboat.

▶ They understand that nobody has a right to survive, that survival must be fought for.

▶ They never give up in a survival situation.

Some of the items in the foregoing list are of a technical nature or involve seamanship. Other literature more appropriately addresses those topics. In this chapter we refer to such matters only as they relate to the maintenance of the physiological status quo or to give an overall flavor of the nature of the potential problems. We direct the following advice primarily toward survivors in inflatable life rafts because they are currently the most frequently used survival aid. Regrettably, life rafts may prove to be of little value if survivors do not understand how to get the best out of them. Having said this, the principles described here also apply to survivors in lifeboats and other survival craft. Regardless of the type of survival craft available, the first and most essential step is to formulate an overall survival strategy.

SURVIVAL STRATEGY

The Baileys (1974) were exemplary in the manner in which they calmly prepared for abandoning their yacht: They made the most of the time available to salvage essential items. They demonstrated that they were aware of the potential problems they faced. Regrettably, we cannot say the same of others.

Disaster on Board a Fishing Boat

A 15-meter (48-foot) steel-hulled commercial fishing boat sailing off the coast of Hawaii in December 1996 unexpectedly experienced a hard thump against the port side of the hull, which caused it to veer off course. Before the man on watch on the bridge realized what had happened, the bilge alarm sounded, signaling the in-rush of water. He immediately started the bilge pump, but the rate of ingress of water flooded the batteries within 90 seconds. Realizing the imminent danger, he rushed forward to wake the other occupant of the vessel, who was resting in the bunk space after dislocating his shoulder the previous day in bad weather. As by now the boat was listing heavily, the first man then proceeded to launch and inflate the six-man life raft. As he turned to retrieve some additional stores from the galley, he realized that the boat was sinking fast so he grabbed the EPIRB (emergency position radio indicating beacon) and a backpack. Assisting his injured colleague, the two men scrambled on board the raft just before the boat sank. The interval between the collision and the boat's sinking was about five minutes.

   The men activated the EPIRB, but it flickered for only a moment before going dead. Still, they felt it had signaled long enough to alert the Coast Guard, so they opened their survival rations and had a party! The backpack contained a Nintendo Game Boy with two games, a spare set of batteries, a

(continued)

(continued)

Walkman, an electric razor, a paperback book, a screwdriver, a mini-blowtorch, and several packs of cigarettes!

In the succeeding days before they were rescued, the men experienced severe cold at night, stormy seas, periods of calm, and tropical heat. Fortunately, they were able to capture sufficient rainwater to meet their requirements, but just barely, and they caught a few fish to sustain themselves.

The absence of communication from the boat for five days precipitated an air–sea search by the U.S. Coast Guard. After a further five days of searching over an area of more than 200,000 square miles, authorities abandoned the search. Fortunately, another tuna-fishing boat happened upon the men and rescued them after 28 days adrift.

Statements of denial such as "It will never happen to me" or "We only go a few miles offshore" are a sure recipe for disaster. Every year, many more leisure sailors and fishers die a few miles offshore than ever die in deep blue water activities. All boat owners need a survival strategy, regardless of where they are operating. Such a strategy will vary depending on the nature of the activity. Consideration must also be given to possible sea states, water temperatures, weather, age and fitness of the crew (passengers), proximity of rescue facilities, and so on. If you are a boat owner or captain, undertake a risk analysis and formulate a plan to contend with the identified risks. If in doubt ask the advice of an experienced authority.

When formulating your strategy, do so under the following headings:

▸ Actions to take before abandoning
▸ Actions to take during abandoning
▸ Actions to take after abandoning

Obviously, time and circumstances will play an important role in how you implement your plans, but the strategy should allow for such eventualities.

Measures taken in the precious minutes before abandoning may save lives in the succeeding minutes, days, or weeks. Don't wait for the event to occur before thinking about what to do; have your contingency emergency plans worked out for various eventualities. Planning is more than just preparing the emergency grab bag; that should just be the starting point.

What would you do if you had only a few minutes? What would you do if you had 5 minutes, 10 minutes, or more to prepare to abandon? It is easier to think and plan in the cold light of day than in a crisis. Imagine trying to formulate a survival strategy when you are struggling to save the boat, in the dark, while trying to calm and direct your panic-stricken crew or passengers. When the dreadful reality suddenly hits and you realize all you can do is abandon, you don't want your only plan to be simply launching the raft and collecting your grab bag.

Having thought through these scenarios, you will probably revise existing stowage plans so that some items are easily accessible in the event of an emergency rather than positioned for the convenience of everyday use. Stowing the life raft or other crucial survival aids in a sail locker buried beneath piles of sails is foolish. After all, everyone understands that it would be unwise to store a fire extinguisher at the back of a food locker simply because it is rarely used. Using the same logic, the grab bag and supplementary equipment and clothing should be within easy reach of the most likely point of escape from the vessel in the event of an emergency.

Valise-packed life rafts should be stowed in a dedicated weatherproof locker easily accessible to the cockpit. Rafts packed in special weatherproof containers can be stacked in secure locations on the open deck or attached to the transom. Whichever type of raft you chose, it must be regularly serviced. Ideally, you should seek the advice of the manufacturer or maintainer on the suitability of the chosen location.

Bear in mind the possibility the vessel may capsize when you are below in the cabin. When you subsequently make your escape, will you be able to access the survival aids? Is some of the survival equipment stowed in an area likely to be damaged by a fire? If so, it will be of little use if the fire gets out of hand and you are forced to abandon. You should think through such scenarios before deciding on the final disposition of your survival equipment and clothing. Make sure that others know where the equipment is stored. You may be the one who falls overboard or becomes otherwise incapacitated.

The Zeus III

On 27 September 2000, a Greek chartered schooner, *Zeus III*, with a party of 32 elderly American tourists on board, struck a submerged rock and sank in about 15 minutes. After his rescue, a Mr. Jack Corn, from Boston, told a Greek television reporter that the crew appeared ignorant about basic evacuation procedures: "They didn't know how to open the canisters the life rafts were in." The life jackets were buried beneath boxes of toilet paper.

Unfortunately, in that example one passenger lost his life—that was one too many for an incident that was eminently survivable. The incident serves as a timely reminder of the requirement for all crew members to be thoroughly conversant with the safety equipment and for all passengers to have a comprehensive safety brief.

SAFETY BRIEF

A safety brief should become a standard operating procedure as soon as new crew or passengers join the vessel, even for a one-day cruise. This safety brief, which may involve demonstrations, should cover the positioning of the safety equipment and how it is used (including fire extinguishers). For yachts and other small craft it should also describe:

▸ The correct way of donning and securing the type of life jacket on board (this must be demonstrated and practiced).

▸ The importance of using a personal safety harness when on the open deck in any weather.

▸ Man-overboard procedures (display on a readily accessible "idiot" card).

▸ The use of the VHF in making Mayday calls. A Mayday prompt card should be stuck in a prominent position adjacent to the VHF set. Everyone should be instructed on how to switch the set on, select the distress channel, and transmit a Mayday. All should know the location of the EPIRB and how to activate it.

▸ The allocation of specific emergency tasks to each crew member. Each should be responsible for collecting some item or items before abandoning. This procedure lessens the chance that something vital will be forgotten in the heat of the moment before abandonment.

Although some might regard it as a humorous interlude, the briefing will remind all that the sea can be a dangerous place. More important, it might save lives. Finally, should an accident occur that results in the loss of life, the failure to give such a briefing could prove costly in litigation.

PREPARATION FOR ABANDONING

While some are working hard to keep the vessel afloat, you should delegate others to prepare for abandonment should the need arise. These tasks will include obtaining the EPIRB and transmitting a Pan or Mayday call. You should initiate an early call for help, as soon as it appears that there is a serious problem. An early warning to the rescue authorities of an imminent problem in their area could make the difference between saving life and simply recovering dead bodies. It is easy to recall a helicopter or rescue craft if those on the vessel can bring the situation under control. But if you have not transmitted an early call and the situation suddenly deteriorates, you may not have time to send a Mayday or you may be rushed. Calls sent at the last minute may be scrambled or omit some vital piece of information. Survivors will then be exposed much longer to an adverse environment before rescuers arrive. In a shipping lane or near the coast, fire a parachute flare downwind at an angle of about 15 degrees from vertical to attain optimal trajectory. In ideal conditions the flare can be seen from 64 kilometers (40 miles) away.

Don, or make available, as much warm clothing as possible. Do not forget gloves and head covering (a spare set of warm socks could prove to be a real luxury). If possible have a waterproof and windproof outer layer or, even better, a survival suit.

The Survival Suit Factor

In June 1998 the South African fishing vessel *Sudur Havid* got into difficulties and sank in the South Atlantic, off South Georgia, in force 7 to force 8 winds and 9-meter (30-foot) seas. "The ship began to take in water just before 1300; a distress call was transmitted and about 1600 the order was given to abandon. The sea temperature was around 1 degree Celsius (33 degrees Fahrenheit). Before I boarded the life raft I rushed back and got my survival suit. This was my own suit that I had brought with me. Only the Captain and I had 'survival suits.' I got back to the life raft just before the ship sank. Unfortunately the raft had shipped a lot of water and we were nearly waist deep in near freezing water. After a very short time people were moaning with pain and the cold began to numb their limbs. Not long after some of the crew began to die." A few hours later another fishing boat arrived on the scene in answer to the distress call. Only 7 of the 17 people in the raft were recovered alive (Mathew Lewis, Aberdeen, Scotland, 1999).

Ensure that everyone is wearing a properly secured life jacket. A loose-fitting life jacket may ride up and come off in the water or cause the wearer to float in a dangerous attitude, and thus be a liability rather than an aid. If time permits, it is worth instituting a buddy check to ensure that all buckles and ties are secure and that harnesses are tight. As early as possible, issue everyone tablets to counter seasickness. The body needs time to absorb the drug so that it can commence working before people enter the nausea-inducing environment of the life raft.

Release the life raft from its stowage place and, if necessary, push it to the optimal launching point. Take extra care not to slip or trip on the rolling deck and drop the raft over the side before you secure it. In ships, rafts are usually stowed in racks on sloping surfaces from which they can be released to roll into the sea under the influence of gravity (this may not occur if the vessel is listing). In the event that the ship sinks before the rafts are released, they will float automatically to the surface once the hydrostatic pressure-release mechanism is activated. Before launching the raft, be sure that the end of the painter is securely attached to a strong fixture. Avoid launching the raft earlier than necessary. Chaffing against the side of the vessel may damage the raft, or a sharp object may puncture it. In some situations, such as a fire that looks as if it is getting out of control, it may be better to launch the raft early to enable it to inflate fully, provided the risk of damage is slight. Early launching also facilitates the loading of supplementary aids and stores. Secure the raft in a position where it is least likely to be damaged.

Usually, it is preferable to launch the life raft in the lee of the vessel. The hull will provide some protection from the weather, and the distance from the deck to the raft may facilitate boarding without having to enter the water. After launching the container, pull on the painter to inflate the raft. There is usually about 8 meters (25 feet) of excess painter to pull before it reaches the point of attachment to the firing pin of the gas cylinder.[2] Should inflation not occur when all the surplus length of the painter is withdrawn from the container, give a final snatch pull on the painter to trigger it. Once the raft starts inflating, try to keep it close to the lee of the boat. In windy conditions the aerodynamic force on an empty, fully inflated raft will be difficult to control.

Full inflation normally takes about 60 seconds, although it may take longer in cold climates and can sometimes be incomplete. If possible, allow time for the raft, particularly the canopy arches, to inflate fully before boarding. The additional pressure caused by body weight may cause gas to escape through the pressure-relief valves before full inflation has occurred. In hot climates, full inflation will occur much more quickly, in about 30 seconds, and you may hear excess gas leaking from the pressure-relief valves in the buoyancy chamber. If the raft inflates upside down, try to right it before boarding. All sea-survival programs teach the technique, so you should be able to do it. In windy conditions, the raft will have a tendency to blow over and invert after it has inflated. Consequently, it is wise to have one or two of the heaviest, fittest members of the crew board the raft early to provide additional stability and to ensure that the ballast water pockets are filling. Resecure the painter so that the raft remains adjacent to the parent vessel but be ready to cut

---

2. The inflating gas is usually either carbon dioxide (CO_2) or nitrogen (N). In very cold environments, it is preferable to use nitrogen because carbon dioxide doesn't expand as well and will produce dry ice that may freeze and block the inflating mechanism.

it in a hurry if necessary. If possible, the person or persons in the raft should make preliminary checks for leaks and start pumping the buoyancy tube if inflation is incomplete. At this stage, if time permits and the sea state is such that inversion of the raft appears unlikely, load the additional stores, grab bags, and so forth. Attach any emergency water containers to the external lifelines on the outer rim of the buoyancy chamber.

Before finally abandoning your vessel, ask yourself again if you have done everything you possibly can to prevent its sinking. In some yachts, emptying the water tanks will provide additional buoyancy that can help maintain flotation and thus present a bigger target for rescuers to locate. Remember that your boat is the best lifeboat as long as it remains afloat.

Finally, make sure the EPIRB is on board the raft. If you can, bring a hand-held VHF and GPS to facilitate communication with the rescue services and provide an accurate location. In some regions adjacent to the coast, mobile phones may prove useful, but you should not depend on them.

ABANDONING

When you finally make the decision to abandon, avoid entering the water if possible. Even when wearing a survival suit, it is much more difficult to enter the raft from the water. The effort can be exhausting for many and impossible for some without assistance. Furthermore, in the process of boarding a raft from the sea, a person will bring aboard a quantity of water. Avoid jumping on top of the raft if possible. If jumping is necessary, do so only into the entrance, and not from a height of more than about two meters (six and a half feet). Jumping on the canopy may injure someone concealed beneath it, or you may damage your legs on the hard survival pack inside the raft. If entering the water is unavoidable, do so slowly, if possible, to reduce cold shock. If abandoning from a high-sided vessel and davit launching of boats or rafts is not possible, get as near to the water as you can before attempting to jump. Climb down a net, ladder, or even a fire hose rather than jump from a height.

RECOVERY OF FELLOW SURVIVORS

Although the objective should be for everyone to board the raft dry, this will not always be possible. In such circumstances, those who board first must help the remainder. Recovering other survivors from the water requires some skill and strength. These skills, and the vital actions one should take on boarding the survival craft, are best learned in a properly supervised program conducted by an approved survival-training organization. Even in such ideal conditions, one can readily appreciate the difficulties involved.

With the recovery of each survivor, it is common for a significant volume of water to enter the raft, adding to that which may already have entered through wave action. If you recover unconscious casualties, be sure not to leave them lying facedown in this water while you help the next victim from the sea. In fact, it is better to recover most of the fit people first. They can distribute themselves about the raft to assist with stability and provide basic first aid to any casualties coming on board. You should then recover the less able, leaving one or two fit people in the water to assist in moving them to the entrance and to prevent them from drifting off. Finally, the last person in the water should quickly check around the raft to ensure that no incapacitated stragglers are hanging on to the far side. Once all are on board a quick head count can confirm whether anyone is missing.

During the boarding procedure people tend to gather around the entrance, either endeavoring to help others or simply recovering their breath by sitting close to where they boarded. Others remain near the entrance because they find the interior of a raft claustrophobic and are afraid that if it inverts they may not be able to escape. Those prone to motion sickness are also likely to stay near the entrance to give themselves a visual reference in the hope of lessening nausea. Regardless, it is important for someone to take firm control early and ensure that the weight is evenly distributed about the raft to enhance stability. Once you can account for everyone, cast off.

AFTER ABANDONING

The words *cut, stream, close,* and *maintain* should be engraved in the memory of every participant of a sea survival-training program. When these concepts become second nature, they can save lives.

Cut refers to cutting the painter to set the raft adrift. Once free of the sinking vessel, it is advisable to stand off. You will not be "sucked down with the sinking vessel," but the raft may become entangled in the rigging as the vessel sinks. You may find it necessary to push the raft along the side of the drifting, sinking vessel to clear the lee.

Once clear of the lee, *stream* the sea anchor. The anchor will provide valuable additional stability as well as slow the rate of drift, thus assisting rescuers' attempts to locate you. Before *closing* the canopy entrance and any other apertures to retain heat, take a final look around to make sure that no one remains in the water. At the same time check for potentially useful flotsam (seat cushions, etc.) drifting nearby. When closing the entrances use slipknots so that fingers that may be numb from cold can untie them quickly and easily.

If a significant amount of water is swilling around the raft, you should bail it out downwind before closing the entrances. Most life-raft bailers hold only a small volume of water and are frustrating to use if a large volume must be removed. Use common sense and whatever is at hand to cope with such volumes (for example, the portable bilge pump from your grab bag or a sailing boot, often the most effective bailer for getting rid of a large volume of water quickly). Once you are down to residual levels, the raft bailer becomes useful. Finally, use the sponge to mop up the drops.

When the raft interior is reasonably dry, take stock of the situation, check for damage, assess your priorities, open the life-raft survival pack (appendix 9.1), and make a quick general inventory of the equipment and stores available. In very cold conditions make sure to perform vital actions requiring manipulative skills early, before the cold impairs manual function. These tasks include opening the plastic wrappings of flares and other items in the survival pack. Make sure you stow these securely in a pocket or tie them so that a wave passing through the raft or a capsize won't wash them away. Don't forget to secure the inflation pump (bellows); its loss could be disastrous in the longer term.

When time and other lifesaving priorities permit, perform necessary first aid on any injured casualties. Keep it simple but effective until time permits a thorough appraisal of the condition and careful attention to the injury. For example, a simple pad applied to a bleeding wound and held, or crudely tied in position, will help stem bleeding until someone can cleanse and properly dress the wound. Obviously, an unconscious victim will require attention at a much earlier stage to ensure that the airway is clear and maintained.

Commence the *maintenance* procedures, which should form part of the watch-keeping duties from now until rescue. Check for leaks and the state of inflation of the buoyancy chambers. The chambers will need periodic topping off to replace the air intermittently escaping through the pressure-relief valves (day-to-night differences in temperature will affect the contained volume of gas). Bail out any excess water and use one of the sponges to mop up the remnants. Keep the second sponge dry and uncontaminated by salt water if possible so that you can use it to collect condensation for drinking water (see chapter 8). Unless your raft is fitted with an internal bailer, you will have to open a portion of the entrance to discard the bailed water. Loss of some of the internal heat will thus occur, but in the longer term having a dry raft will conserve more heat. Achieving complete dryness in a survival situation is almost impossible. The depression in the floor created by the weight of bodies will nearly always retain some water.

Staying Warm, Staying Alive [Thermal Insulation]

Having assured yourself that the raft is as seaworthy as you can make it, the next priority is thermal insulation. Using the information provided in chapters 2 and 3 on thermoregulation, implement a strategy to help maintain thermal balance. If possible, establish an order of priority to eliminate the biggest source of heat loss or gain first and continue thereafter in descending order of priority. Practical considerations, however, often dictate the order in which you take measures. For example, it is obvious that if the buoyancy tube in an inflatable raft is only partially inflated and waves are continually breaking inboard, there is little point in wasting energy bailing until you control the ingress of water. This may involve closing the windward entrance, fixing a leak, or simply pumping more air into the buoyancy tube. In very cold conditions, the expansion of gas may be insufficient to fill the buoyancy chambers satisfactorily, so they will need to be topped off. Similarly, if several inches of free water are swilling around the floor of the raft, rapidly conducting body heat to the sea beneath, there is little point in fussing about closing canopy apertures to control convective heat loss.

Death From Hypothermia

"The woman, he reported, was becoming more and more listless and limp. She occasionally slid down into the water in the raft and was pulled up by others, who tried to massage her and shake her to arouse her. One hour prior to their rescue, she died" (MV *Estonia,* Final Report 1997).

If by partly closing the aperture you can stem the inflow of water, then do so while continuing to bail. Many current larger rafts have an effective means of internal bailing through a system that incorporates a non-return valve built into the floor of the raft. This enables bailing without opening the apertures.

Life-raft manufacturers recommend inflating the floor to reduce conductive heat loss to the sea. In terms of life-raft husbandry, however, this is a relatively low-priority task, not because it is unimportant but because it is almost unachievable. In practice it is impossible to exert sufficient pressure with a hand pump to counter the downward pressure from the weight of the seated body. Therefore, the result of the hard work of pumping is a ballooning of the floor between the seated occupants while the floor beneath them remains airless and compressed. Furthermore, any free water in the raft will accumulate in these depressions in the floor, thereby intensifying conductive heat transfer. In the short term, once you are confident that the risk of capsize is small, it is better to take off your life jacket and use it as an insulating cushion to stem heat loss from this source (figure 9.3).

Provided a good canopy seal can be maintained, the heat given off by the occupants will help warm the environment within the raft, thus reducing convective heat loss. If everyone is wearing survival suits or waterproof clothing, however, less body heat will escape to warm the atmosphere. Clothing and head coverings will help reduce heat loss through radiation.

Those in wet clothing should remove the outer layers, squeeze them dry, and put them back on. Some body heat will be lost in drying the clothing, but by now the immediate cold threat will have passed. The body should be able to defend this minor thermal challenge with activity or shivering. For those not wearing specialized protective clothing, evaporation will be the main source of body heat loss; this will continue

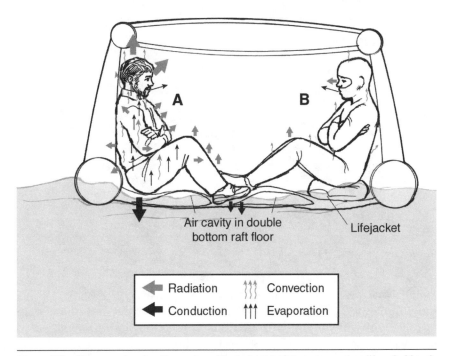

Figure 9.3 Diagram showing the routes of body heat loss in two men in a life raft. Man A, seated on the raft floor in a small pool of water, is wearing ordinary clothing that is wet. Man B is seated on his life jacket and wearing a survival suit with an integrated hood. Beneath the survival suit he is wearing well-insulated undergarments. He has also folded his arms close to his body and put his hands beneath his armpits.

as long as clothes are warm and wet and the environmental water vapor pressure is less than that at the surface of the clothing. If the internal environment of the raft can be well contained by sealing the apertures, it will warm up reasonably quickly and the water vapor pressure will increase, thereby reducing evaporative heat loss. But good seals are often difficult to make and maintain in these circumstances, and thus cooling through evaporation is likely to continue. Should people begin to shiver, consideration should be given to using individual thermal protective aids (TPAs). These need to be robust and capable of withstanding the wear and tear likely to occur in a raft. Such quality is not always available: "One of the passengers found a thermal protective suit and tried to put it on. The suit was too thin and tore in several places" (MV *Estonia*. Final Report 1997).

Some TPAs are little more than large polyethylene bags (figure 9.4) that the individual can climb into and secure around the head, leaving

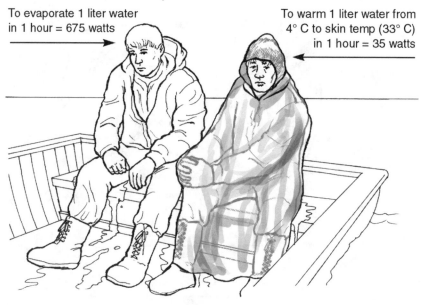

Evaporative heat loss

To evaporate 1 liter water in 1 hour = 675 watts

To warm 1 liter water from 4° C to skin temp (33° C) in 1 hour = 35 watts

Figure 9.4   The detrimental effect of evaporative heat loss in two survivors. Both had saturated clothing on rescue. After rescue, one is immediately enclosed in a polyethylene bag. Once water vapor saturates the air in the bag, the heat loss he experiences is predominantly that required to warm the wet clothing to skin temperature. The other casualty, however, will continue to lose heat through evaporation. Given that maximum shivering is capable of producing around 500 watts, he will quickly run into heat debt.

just the face exposed. Within the bag, water vapor quickly saturates the atmosphere so that further evaporative heat loss stops.

An effective TPA will reduce heat loss and enable the body to replenish its heat stores through shivering. Comfort will improve temporarily, but eventually the loss of insulation in the clothing through saturation from moisture vapor will result in intermittent bursts of shivering. Therefore, when conditions are favorable and the raft is reasonably dry, it is better to remove the TPA and permit the damp clothing to dry. But if shivering continues, it is advisable to put the TPA on again. This may be necessary at night, if the number of occupants is too few to generate sufficient heat to warm the internal environment, or if a good canopy seal is unattainable.

With protracted survival in cold conditions and insufficient rations to meet the extra energy demands, shivering intensity will diminish. With starvation, it may be totally absent. Therefore, deep body temperature will fall. The risk of cold injury significantly increases in this situation (chapter 10), particularly if dehydration is also present.

Leadership

With everything secured and the raft made shipshape, you should follow standard leadership practices.

- ▷ Keep everyone involved and fully briefed on his or her duties.
- ▷ List the dos and don'ts of living in a raft. Topics include stability, sharp objects, use of flares, actions to take in the event of puncture, rain, water and ration discipline, fluid and energy conservation, and so forth.
- ▷ With a naive crew, it is often safer to confiscate all sharp knifes to reduce the likelihood of inadvertent puncture.
- ▷ Because conditions are frequently cramped and tempers rise when people attempt to stretch to ease limb discomfort, discuss this potential problem early to warn people of the need for tolerance.
- ▷ Steer conversation away from the cause of the disaster if possible and avoid placing blame; doing so can damage morale.
- ▷ Be prospective in your thinking rather than retrospective.
- ▷ Be positive and don't even consider that you will not be rescued.
- ▷ Make plans for the days ahead. A first step in the right direction is to organize the day, establish a watch-keeping roster, and brief watch keepers on their duties.
- ▷ Try to make everyone feel that he or she is, or will be, playing an essential role for the common good.
- ▷ Finally, and above all, maintain discipline.

We discuss some further important psychological considerations in chapter 10.

Duties of Watch Keepers

The immediate objective is to stay alive. To this end, continued good raft husbandry is essential. Next is an awareness of the need to help potential rescuers locate you. Check that the EPIRB is switched on and the VHF is ready for use. Conserve the battery life of lights by switching them off during daylight hours. Small life-jacket lights may not appear to be very bright, but at night they stand out against the black background of the sea. To searchers wearing night goggles (image-intensifying equipment), they can be almost blinding. If you see or hear a boat or search aircraft, don't fire all the flares at once in a mammoth firework display. Formulate a policy of when and how to use the flares. Brief the crew also on the use of other location aids. When you sight potential rescuers make sure that everyone does not rush to the entrance to catch sight of them. Likewise, don't waste precious water rations on a celebratory drink. The would-be rescuers may not see you or may be unable to rescue you for some reason (see the Poon Lim story earlier in this chapter).

Keep a lookout for useful flotsam (empty water bottles, bits of wood, and so on). You may be able to put to good use items that others have discarded or items that have floated off your vessel as it sank. If necessary, you can tie some items to the outside of the raft provided they don't upset its sea-keeping abilities. In the Tropics keep a lookout for gathering rain clouds or the occasional turtle.

Lookouts may need additional protective clothing in adverse conditions. Such matters and the time they spend with their heads outside the canopy will have to be determined in light of the prevailing conditions. Lookouts should try to maintain the thermal integrity of the ambient environment of the raft while on duty. In the Tropics watch keepers should protect against sunburn and remain in shade if possible to conserve water (see chapter 8).

In crowded rafts it may be necessary to ventilate the raft periodically to control the buildup of carbon dioxide in the atmosphere from expired breath. Headaches are usually the first indication of such a buildup, although they will also occur with dehydration. Because carbon dioxide is heavier than air, the buildup will be more noticeable near the floor of the raft below the level of the buoyancy tubes. The constant buckling of the raft in a seaway coupled with the difficulty of achieving an airtight seal in the canopy tempers this threat. Still, you should consider it if some of the occupants complain of headaches, especially in calm weather. A few minutes of ventilation every hour will clear the atmosphere. Should an inboard leak occur from the buoyancy tube of a raft inflated with carbon dioxide, the threat is more serious. After you have mended the leak, vigorous ventilation is in order.

Seasickness

Seasickness can be debilitating to occupants of survival craft. Chapter 10 discusses it in detail.

Saltwater Ulcers

Another source of debilitation are painful ulcers of the skin (see also chapter 10). These frequently occur in survivors after they have been adrift for some time. Ulcers are usually the consequence of several contributory factors:

▶ Poor local circulation through dependency and inactivity

▶ Softening of the skin from water logging

▶ Accumulation of salt crystals, which abrade the soft, waterlogged skin over the pressure points—seat, elbows, and heels

▶ Progressive loss of muscle bulk in starvation conditions, which reduces the cushioning between the skeleton and the skin in pressure areas

▶ Protein deficiency, which delays healing

The small abrasions that initially appear are slow to heal because it is difficult for a scab to form or remain in position long enough for the healing tissue beneath to take hold and develop. Concurrent starvation with protein deficiency will further delay healing or even inhibit it completely. Consequently, secondary infection and further growth in the dimensions of the ulcer are likely. In the end ulcers can be extremely painful and debilitating.

Thus raft occupants should undertake a procedure of skin care early on as a matter of routine:

▶ Try to avoid cutting or abrading the skin.

▶ Keep the skin as dry as possible, although this is often difficult in these circumstances.

▶ Gently brush the salt crystals from the skin at frequent intervals.

▶ Wash the skin in these areas if sufficient rainwater is available (spare, nonpalatable water is ideal for this).

▶ Several times a day elevate for a short period the body parts that are normally dependent.

▸ Expose the ulcers to air currents but avoid exposure to direct sunlight or spray.

▸ Gently exercise the limbs to promote circulation if adequate rations, space, and general conditions permit.

▸ Finally, apply some form of emollient cream, such as Sudocrem or a similar substance, to the skin over the pressure points.

Once the ulcers have developed they are almost impossible to treat in a survival situation, but try to elevate the affected areas and keep them dry.

Water and Food

Chapter 8 provides advice about both water and food.

Urination and Bowel Movements

Some people have difficulty urinating when in a life raft. A side effect of motion-sickness medication may be part of the cause (hyoscine; see chapter 10). It is advisable to urinate a small volume early, when the inclination first arises, rather than delay as long as possible lest you have difficulty at that stage. Holding on to a full bladder of urine offers no advantage to body water balance; water cannot be reabsorbed from the bladder. By the second day, urine output will only be about 500 milliliters (17 ounces) per day. Thereafter its volume will depend on the fluid and food rations being consumed. If these are restricted, urine volume will fall below 300 milliliters (10 ounces) per day, and the urine will be concentrated and dark.

The first bowel movement is also a difficult psychological hurdle, but thereafter it is probable that significant bowel movement will not occur until intake of bulky food recommences. No harm will come from not producing bowel movements. They are unnecessary when a minimal amount of food is being eaten and all the fluid consumed is being absorbed. On restoration of a normal diet after rescue, bowel activity quickly returns to normal. Those who have swallowed a lot of salt water before boarding the raft may initially suffer diarrhea, which will intensify their dehydration.

DEATH

The possibility exists that survivors will die in the raft, so you should formulate a policy about how to deal with such a situation. If rescue is likely within a few days (if you are close to shore, for example), it is advisable to retain the body for legal purposes and out of consideration for the next of kin. In remote areas of the world, where the probability of early rescue is low, you should consider burial at sea. The continued presence of a dead colleague in the raft will be disturbing and will damage morale. Before burial, however, salvage the victim's clothing to supplement that of the remaining survivors. Retain the personal effects of the deceased for the next of kin.

OPEN BOATS

The principles of long-term survival for those in life rafts, described in the preceding paragraphs, are also valid for those in open boats. Although the rate of heat loss by conduction may be less in a boat than in a raft, convective, evaporative, and radiant heat loss may be considerably greater unless the boat is enclosed (as with a TEMPSC; see next section), or an effective awning can be rigged. Adequate protective clothing is therefore essential, including some covering for the high heat-loss area of the head.

TEMPSC

Totally enclosed motor propelled survival craft (TEMPSC) are increasingly being employed on offshore oil and gas platforms and large ocean-going vessels. They provide excellent protection from fire and most environmental conditions. One drawback is that occupants may be more likely to suffer from seasickness than those in rafts.

RESCUE

Remember the adage that "no one is a survivor until he or she has been rescued and made a full recovery." The arrival of a search aircraft overhead does not mean that rescue is imminent. It simply means that you have been located. Rescue depends on the proximity of potential rescue craft, weather, problems of being located by rescue craft other than an aircraft, flying conditions at the base from which the search aircraft are operating, aircraft serviceability, and many other factors. These circumstances could mean that days may pass before you can be rescued. So although it may be a time for a thanksgiving prayer if you are so inclined, it is not time for a party. You must maintain survival discipline until you are safely rescued. Discipline requires strong leadership, and ample historic precedent justifies such a policy.

PUTTING IT ALL TOGETHER

We finish this chapter with an account of a survival incident that provides a stark contrast to the incident quoted at the beginning of this chapter. It illustrates the importance of understanding the principles of survival, being resilient enough to overcome adversity, and having the will to survive.

The Survival of Tony Bullimore

Tony Bullimore used his understanding of the nature of the threat he was confronting when his 18-meter (60-foot) yacht suddenly capsized after losing its keel in a Southern Ocean storm 2,736 kilometers (1,700 miles) southwest of Perth, Western Australia, in January 1997. The capsize occurred during the Vendee Globe round-the-world race only two weeks after the dramatic rescue of the French yachtsman Raphael Dinelli by Peter Goss and on the same day as the rescue, by the Australian navy, of Thierry Dubois, who had also capsized. The 56-year-old Bullimore found himself trapped inside the inverted yacht, floating in 12- to 15-meter (39- to 49-foot) waves with a 60-knot wind and freezing conditions. Fortunately, he was wearing an expensive yachting dry suit similar to those used by many current ocean-going sailors. Standing on the cabin roof, waist deep in cold water inside the 2-by-3-meter ($6\frac{1}{2}$-by-10-foot) cabin, he took stock of his situation. Should he abandon or stay put?

The yacht had a number of airtight compartments running the length of the hull, which had not been damaged in the capsize. Without the weight of the keel, the boat was in no immediate danger of sinking. With each wave, both water and air gushed in and out through a broken forward cabin window, like a miniature Niagara Falls. Having assured himself that he had sufficient fresh air to breathe, he quickly realized that he had to get out of the water to conserve heat. This he did by cleverly constructing a hammock, above the water line, using a cargo net (see figure 1.2 on page 10). This was no mean feat because he did it in total darkness, as he could not find a flashlight. Well outside the immediate range of Australian rescue capability, Bullimore knew that the search for him would commence once his EPIRB, which he had activated, was detected, but he also knew that some days would pass before rescue could occur. Concerned that after several days the crew of a search aircraft might interpret the sight of the upturned hull wallowing in the big seas without any sign of life as evidence that he had perished, he decided that he had to release the life raft and secure it to the hull to show he had survived the capsize and was probably alive inside the hull. This involved diving below the water to exit through the submerged door to the partially flooded control compartment, and then through another door to the cockpit. After about a dozen attempts, each involving breath holding and working underwater with numb fingers—he had lost the tip of one earlier in the knockdown—he eventually released the raft but couldn't free it from beneath the hull. Frustrated, he returned to the cabin and settled in his hammock to await rescue. Each of his forays into the water sapped his body heat, making it impossible for him to spend more than a few minutes at a time outside the cabin. About two hours of intense shivering back in the cabin followed each attempt. After sufficient recovery he forced himself outside again.

In the succeeding three days, before eventually being rescued by an Australian navy ship, Bullimore sustained himself with a limited supply of water and some chocolate. On rescue, he was hypothermic, dehydrated, and suffering from cold injuries. The world's media said he was lucky to be alive. We disagree. He survived because of several factors, not least of which were his personal training and knowledge of his survival requirements (he had been a Royal Marine officer) and the professionalism of the Australian rescue personnel. He analyzed his situation correctly, made the right decisions, and optimized the use of available equipment. He proved again that those with the greatest chance of survival have the correct equipment, know how to use it, have a good understanding of the threats, and know how to improvise to counter them. Such knowledge comes from wide-ranging forethought, training, and experience.

Chapter Summary and Recommendations

> The two great physiological threats to survivors in life craft are cold and insufficient drinking water. The ability to deal with these comes from understanding the physiological principles covered in this chapter and acquiring appropriate seamanship and technical skills (covered in other publications). We highly recommend undertaking a sea-survival program.

> Formulate a survival strategy under the following headings:

  a. Actions to take before abandoning

  b. Actions to take during abandoning

  c. Actions to take after abandoning

▷ Conduct a safety brief for crew and passengers before sailing.

▷ Ensure that your emergency grab bag has the appropriate contents for the environment in which you are operating. Have a reserve list of supplementary items to take if time permits.

▷ An early Pan or Mayday call is important.

▷ As long as it remains afloat, your boat or ship is the best lifeboat.

▷ Activate the EPIRB before boarding the raft in case you lose it on boarding.

▷ Avoid entering the water if possible.

▷ Some fit people should board the raft first.

▷ Follow standard leadership practices.

▷ Take antiseasickness tablets early.

▷ Cut, stream, close, maintain.

▷ Stay warm and stay alive.

▷ No one is a survivor until he or she has been rescued and has made a full recovery. Maintain survival discipline to the end.

Appendix 9.1
LIFE RAFT EQUIPMENT

Most life-raft containers include packs of survival aids supplied by the manufacturer and enclosed inside the packed raft. The minimum contents stipulated by the International Maritime Organization (Safety of Life at Sea) for craft with up to a 12-person capacity are as follows:

Bailer

Sponges (×2)

Leak stoppers (×2)

Pump

Repair kit

Buoyant paddles (×2)

Signal card

Instruction leaflets (survival and immediate action)

Flares: red parachute (×2)
         red handheld (×3)

Radar reflector

Water (0.5 liters, or 1 pint, per person)

Graduated drinking vessel

Fishing kit

Flashlight and spare batteries (and bulb)

Antiseasickness tablets (six per person)

Rescue line and quoit

Safety knife

Sea anchor (×2)

First-aid kit

Sick bag (one per person)

Whistle

Buoyant orange smoke (×1)

Heliograph (signaling mirror)

Thermal protective aids (×2)

Non-thirst-provoking rations

Tin opener

Some national authorities also recommend the inclusion of an inflatable radar reflector with telescopic pole.

Appendix 9.2
SUGGESTED GRAB BAG CONTENTS

The contents of the grab bag are optional except in those countries where the national maritime regulatory agency or sailing-race authorities specify the minimum contents. These usually include the following:

Second (spare) sea anchor

Additional first-aid kit

Extra antiseasickness pills

Buoyant smoke signal (×2)

Thermal protective aids

Parachute flares (×2)

Red handheld flares (×3)

Sunscreen and lip salve

Heliograph (signaling mirror)

Radar reflector

In addition one might consider adding the following items:

Personal location beacon

Handheld waterproof VHF

GPS

Repair kit of the type used by garden pool merchants, adhesives that can be applied to wet surfaces

Balaclava with waterproof outer shell

Waterproof warm gloves

Multipurpose knife

Waterproof matches

Small container of Sudocrem or petroleum jelly

Fracture straps (×2)

Spare eyeglasses (if a user)

Book on survival!

Waterproof flashlight with attachment clip

Spare batteries

Cyalume sticks

Blunt-ended heavy-duty scissors

Nylon string

Packet safety pins

Packets of boiled sweets

Medium-sized plastic bags and ties

Small role of kitchen cling film

Small roll of polyethylene refuse bags

Diary (log) and pencils

Additionally, you should consider keeping personal items such as passports, money, travelers checks, other documents that might be useful to have at hand, medication, spare eyeglasses, and so forth in a small waterproof bag. In an emergency, you can quickly throw this bag into the grab bag, ensuring that personal items will not be forgotten.

The grab bag should be waterproof and positively buoyant with a handle that is easy to grab with cold hands. It should have a lanyard or other means of securing it to the body should both hands be required for other purposes. The grab bag must be capable of being stowed safely in a place that is convenient to grab at the last moment before abandoning the vessel. The bag should be checked from time to time to ensure that its contents are not deteriorating and that date stamps of the contents have not expired.

A roll of large, heavy-duty polyethylene refuse bags can have several uses. The bags can serve as makeshift TPAs—one bag to enclose the legs and pelvis and another to cover the torso with a hole cut in the bottom to push the head through. Bags can function as funnels for rainwater and keep spare clothing dry. In an emergency an inflated bag can supplement buoyancy.

In warm climates, the emphasis should be on making additional fresh water available in partially filled containers that will float when thrown overboard (chapter 8). If affordable, a reverse-osmosis pump (chapter 8) could prove lifesaving in remote areas. Several tubes of a strong sunscreen ointment will be valuable.

Additional Items to Consider

If time permits, survivors should salvage a number of items from the sinking boat and load them on the raft or tender or, if the items are bulky and buoyant, float them alongside and attach them to it. In remote parts of the world, where rescue is unlikely to occur quickly, these items may prove extremely useful.

- ▸ Portable bilge pump. The raft bailer is laborious to use and not very effective.
- ▸ Seat cushions to reduce conductive heat loss.
- ▸ Knife and a sharpening implement.
- ▸ Towels and spare clothing.
- ▸ Fenders.
- ▸ Sunglasses.
- ▸ Polyethylene storage boxes (Tupperware) containing carbohydrate foodstuffs—sugar, chocolate, boiled sweets, dried fruit, biscuits (cookies), condensed milk, jams, fruit juice, and so on.

▸ Empty boxes are also useful for other functions.

▸ Fishing equipment or spear gun for use if plenty of water is available and with care to avoid puncturing the raft.

▸ Additional medical stores from the medical kit—skin creams including Sudocrem, and Flamazine for burns, antibiotics, eye lotion, inflatable splints, clear plastic adhesive tape (for wound suturing), antiseptic solution, and spare bandages and dressings.

▸ Camera flash (as location aid).

In summary, survivors should heed the words of Peter Jennings, a Royal Yachting Association yachtmaster examiner: "The grab-bag message is probably that anything may well be useful, so take as much as you can," time and space permitting.

CHAPTER

10

Illnesses, Injuries, and Psychological Trauma

THEY HAD BEEN IN THE OPEN SAILING BOAT for approximately $6\frac{1}{2}$ hours before the capsize, during which the weather had steadily deteriorated. The light rain, which was falling when they first set out, had increased to almost tropical intensity, while the wind increased to force 8. By the time of the capsize the waves had reached almost 4 meters (13 feet). The air temperature was 11.5 degrees Celsius (53 degrees Fahrenheit); the sea 13 degrees Celsius (55 degrees Fahrenheit). The two teenage crew members that ultimately died were both chronically seasick and had been lying on the bottom of the boat under a sheet of canvas and shivering violently for almost 4 hours. In the water immediately after capsizing, both were semiconscious and apparently unable to understand the instructions they were receiving from the man in charge. They were suffering from hypothermia that was ultimately to prove fatal.

This chapter includes information on an assortment of ailments relevant to the topic of survival at sea. These, either alone or in combination with other stressors, can erode morale or decrease survivability.

SEASICKNESS

Seasickness (or motion sickness) is an age-old problem for seafarers and more recently for those traveling in air or space. Seasickness has been the subject of many books and countless papers in scientific journals, many of which relate to travel sickness in general. The reader seeking a detailed review of this work should refer to either the book by Reason and Brand (1975) or the comprehensive review paper by Money (1970). Despite the vast amount of research conducted in this area during and since World War II, most recently by NASA, the search continues for a suitable preventive or curative drug.

In a survival situation, seasickness poses a problem for many people. Its rapid onset and associated misery and discomfort can be devastating at a time when survivors need to be both alert and active to make critical decisions. With time, seasickness can sap both morale and energy. It can also result in a waste of survival rations!

Although the inflatable life raft has saved many lives at sea, for most occupants it is a provocative device for inducing motion sickness. The simultaneous occurrence of several conflicting physiological stimuli causes seasickness. The absence of a good visual horizon at a time when the brain receives intense stimulation from the balance organs in the inner ear produces a conflict of sensory inputs that quickly results in nausea and vomiting. On a large ship, with a high vantage point and open visual reference of a distant, relatively stable horizon (achieved by counterbalancing movements of the head and body), the nausea-inducing sensation from the balance organs is usually overridden. But within the confines of a raft, with its peculiar motion (it twists and turns as it rises and falls with every swell), no stable visual reference is present to counter the central input from the ears. The result is nausea and vomiting, even in habituated sailors. The wallowing motion of an enclosed TEMPSC is also particularly nauseating. All survivors should therefore take some form of appropriate medication (see "Prevention and Treatment" on page 217), preferably before boarding any form of survival craft, because the onset of nausea can be rapid. Furthermore, once vomiting commences, it is difficult to retain further oral medication.

Thermal Implications

Anecdotal descriptions of those suffering from seasickness suggest that it may impair thermoregulation, thus increasing susceptibility to hypothermia. Possible mechanisms may be inactivity, vasomotor changes, and increased heat loss through (nonthermoregulatory) sweating associated with motion illness. Veghte (1972), in a field study of clothing assemblies, exposed three subjects in aircrew life rafts to cold ambient conditions and rough seas. He reported that the subject who consistently suffered motion illness throughout the trial was the only one whose rectal temperature did not stabilize and fell below 35 degrees Celsius (95 degrees Fahrenheit).

Sharpey-Schafer et al. (1958) showed that nausea and vomiting, induced by injections of the drug apomorphine, increase muscle blood flow. Mental stress produces a similar response. Thus both seasickness and mental stress, common in a survival situation, may override heat-conserving vasomotor activity. Descriptions of "cold, clammy, pale skin" in people who are seasick, however, suggest that skin vasoconstriction may be present, but evidence relating to skin blood flow has been lacking.

Mekjavic et al. (2001) investigated the influence of motion sickness on thermoregulation in a cool environment. They conducted an experiment by inducing motion illness through repeated sudden head movements in subjects undergoing G_z acceleration in a centrifuge (a known nausea-inducing stimulus to the balance organs in the ear). After this procedure, and while still nauseous, the subjects were immersed in a tub of cool water (28 degrees Celsius; 82 degrees Fahrenheit), a temperature below the thermoneutral range. Mild motion illness significantly attenuated the vasoconstrictor response to immersion and resulted in a faster fall in deep body temperature compared with that seen with the same subjects in a control condition (G stimulation without the nausea-inducing head movements). This evidence indicates that seasick survivors in cool environments may be particularly susceptible to hypothermia. This effect may be more significant when the motion-illness stimulus is more severe and continues throughout the cold exposure.

Survivors immersed in rough, open water are also prone to seasickness, although less so than those in life rafts in similar sea states.

Prevention and Treatment

Habituation to motion remains the best prevention. Those who can't or don't habituate tend to eliminate themselves from routine exposure by a process of natural selection. But even habituated individuals will occasionally succumb and become sick in a scenario outside their experience, such as in very rough seas or when wallowing while drifting in a swell, particularly when in a life raft.

Anecdotally the most effective "cure" found during the early years of World War II was "Mother Siegal's Soothing Syrup," the active ingredient of which was shown in 1944 to be hyoscine. In the latter years of that war, American research favored the use of antihistamines. In the postwar years, however, several researchers showed that the most effective drug was L-hyoscine hydrobromide. For routine antiseasickness use, the adult oral dosage is 0.3 milligrams, six hourly if required, to a maximum of three doses in 24 hours. The dose for a child is 0.15 to 0.3 milligrams if older than age 10, depending on body size, and 0.075 to 0.15 milligrams for children age 4 to 10. In a survival situation when the adverse side effects (see below) are less important, the adult dose can be increased to 0.6 to 1.0 milligram to a maximum of 2 to 3 milligrams in 24 hours.

A person who regurgitates the original tablet may take a second dose. He or she should try to chew the tablet or retain it beneath the tongue to obtain possible benefit from local absorption in the mouth. Better still is a small adhesive disc or patch containing the drug. This is applied to an area of hairless skin, usually behind the ear. The drug is slowly absorbed across the skin in the succeeding hours. After 24 hours, most occupants will have habituated, to a degree, to the unusual motion. Further medication will be unnecessary unless the weather is constantly changing.

The advantage of hyoscine over other drugs is that it acts rapidly, is more effective at preventing vomiting, and retains its effect for four to six hours after administration (Glaser and McCance 1959). It also has no adverse effect on the rate of heat loss in cold-water immersion (Keatinge and Evans 1960). Hyoscine is the most commonly used active ingredient in many of the popular proprietary brands of antiseasickness cures. Unfortunately, it is not 100 percent effective and has some undesirable side effects that can make it both unpopular and unsuitable for everyday use. These include drowsiness, blurring of vision, and dryness of the mouth. Occasionally, it produces urinary retention and, with repeated larger doses, hallucinations. People with glaucoma should not take the drug.

People generally tolerate antihistamines better, although they also cause drowsiness. Controlled trials, however, have shown antihistamines to be less effective than hyoscine. The less sedating antihistamines, cyclizine or cinnarizine, appear to be the most popular of the antihistamines in the treatment of motion sickness. Drugs commonly used in the treatment of non-motion-related vomiting, such as metoclopramide and the phenothiazines (excepting promothiazine), are ineffective for seasickness.

A number of alternative "cures" for seasickness are available on the open market. These range from a variety of drugs to bracelets/bands, some of which claim to put pressure on relevant acupressure sites! The multiplicity of so-called cures indicates that none of them are particularly effective. Manufacturers can claim success for their particular products largely because of the exceptionally large placebo effect in motion sickness. Most scientifically controlled double-blind clinical trials of drugs for motion illness demonstrate about a 60 percent success for the placebo (control condition)! That said, drugs can produce a greater beneficial effect, and to date, hyoscine hydrobromide remains the drug of choice based on most well-controlled trials.

Seasickness in a Survival Craft

Despite taking seasickness tablets before boarding, several occupants may still feel nauseous. Accordingly, if polyethylene bags are part of the contents of a grab bag, now is the time to distribute them. It is better to contain vomit in a personal bag rather than vomit over others, possibly making them sick too. In addition, mixing the residue with the free water in the partially flooded floor is unpleasant and can trigger vomiting in others. If one is already vomiting, swallowing extra seasickness tablets is pointless because the tablets are unlikely to dissolve sufficiently before being regurgitated. Alternatively, if the person retains all the tablets, they could produce serious side effects. As suggested earlier, if transdermal patches are unavailable, it is worth trying to chew or retain a tablet beneath the tongue. Some of the slowly dissolving tablet may be absorbed though the lining of the mouth. The vomiting of small amounts of bloodstained fluid sometimes accompanies protracted seasickness. Although alarming to some people, this is rarely dangerous and will decline as vomiting subsides with habituation. An alteration of sea state or a burst of activity involving head movement may trigger another bout, so tablets should be provided for susceptible individuals if deterioration in weather is expected.

OSMOTIC DIARRHEA

Survivors who have swallowed large volumes of salt water may suffer from osmotic diarrhea. This condition results from the presence in the bowel of an abnormal volume of fluid with a solute (salt) concentration well above that of body fluid (3.5 percent compared with 0.9 percent). The high concentration of salt in the bowel will attract, through osmosis, additional fluid across the gut wall from the body. This process will increase the overall volume of water in the bowel, although it will tend to dilute its solute content. At the same time, the cells of the intestinal wall attempt to conserve water by pumping it in the opposite direction across the gut wall into the blood. But passive flow by osmosis dwarfs the volume of fluid being transported into the blood stream. The result is an increase in the distension of the gut wall, resulting in contraction of the intestines and expulsion of the contents. The resulting diarrhea rids the body of the undesirable salt but takes fluid with it (see also the section "Seawater Enemas" on page 154 in chapter 8).

OIL CONTAMINATION

Shipwreck survivors are frequently covered in fuel oil. Survivors may even have swallowed or inhaled some. In small quantities, oil is not toxic to the system, although if swallowed it may cause vomiting and if inhaled it may produce aspiration pneumonia. In the eyes it will produce a local inflammation *(conjunctivitis),* which may last several days.

After rescue, the best method of cleansing the oil is to remove the outer clothing and place the survivor under a hot shower (provided it is safe for the survivor to do so; see chapter 11) and gently wipe with cloths or paper towels to remove the excess oil from the skin. Shampoo and bath soap may be used, and specialized skin cleansers, such as Swarfega, will deal with any obdurate patches. You should avoid using solvents and other scouring compounds not intended for use on skin. The eyes may be cleansed with mineral oil (liquid paraffin) followed by a topical

Shipwreck survivors are often covered in fuel oil.

Photograph courtesy of U.S. Coast Guard Public Affairs Staff, Washington, DC 20593-0001.

steroid to relieve the conjunctivitis. When dealing with multiple survivors, initial cleansing from around the mouth, nose, and eyes is all that is required. The remainder of the body can be cleansed later when time permits.

SKIN ULCERS

Ulcers that have developed on pressure areas or because of minor skin damage from abrasions or boils are difficult to treat and will require patience and care. Prevention, along the lines outlined in chapter 9, is the best cure. Once the skin is broken, however, it is notoriously difficult to heal in the damp, salty environment of a life craft. Bacteria found on the skin, which are relatively harmless on intact skin, will grow and multiply in an open wound and delay healing. Concurrent starvation and body-protein deficiency will further delay healing or even inhibit it totally (see chapter 8). Therefore, wounds should be gently washed with fresh water, preferably containing some mild antiseptic. Keeping the wounds dry and elevated will encourage healing. In a life raft, occupants can use some fresh, nonpotable water for cleansing wounds.

Saltwater boils, or pustules, are often the forerunner of ulcers in pressure-point areas and elsewhere. The boils frequently occur in hair follicles because of constant dampness and poor hygiene. In hot weather, the constant warm dampness of the skin beneath damp underwear promotes bacterial growth. It is better to go without underwear in such conditions and, if possible, expose the affected area to a moderate amount of sunlight periodically during the day. Again, washing two or three times a day with fresh water containing mild antiseptic will be helpful. After drying, gently massage some emollient, such as Sudocrem, into the skin over the pressure points. If pus is present, it should be released before cleansing, and the area dotted with iodine or other suitable antiseptic. In severe cases, antibiotic treatment may be required, but the infecting agent will often be resistant to commonly used antibiotics. Pustules developing in areas of chaffing (near wrist seals, for example) are best treated early by removing the cause of the chaffing or at least covering the skin with a bandage or petroleum jelly to reduce the area of friction.

COLD INJURIES

As stated in chapter 2, body cells function optimally at temperatures around normal body temperature (37 degrees Celsius; 98.6 degrees Fahrenheit). When heat loss from superficial tissues exceeds local heat production and the heat gained from deeper organs, tissue temperatures fall.

Because of their high surface area-to-mass ratio, which predisposes them to heat loss, the hands and feet are particularly susceptible in this respect, especially at rest. Tissue temperatures below a threshold level for a period of time will result in cell dysfunction or even damage (cold injury).

The Danger of Immersion Foot

During World War II, 17 men wearing heavy winter clothing boarded a "Carley float" from a ship that was sunk 64 kilometers (40 miles) east of Aberdeen, Scotland. The sea temperature was 2.8 degrees Celsius (37 degrees Fahrenheit), and the air temperature was 0 degrees Celsius (32 degrees Fahrenheit). There was a strong easterly wind and a rough sea. Sitting on the raft with their feet dangling in the water, only 4 of the 17 survived. They were washed ashore after 18 hours. Only 1 of the 4 was capable of climbing a short distance, up a broken cliff, to summon help. All were suffering from hypothermia and "immersion foot." Subsequently, one had some toes amputated, two had long, painful convalescences, and only one was capable of light work ashore.

With body cooling, the temperature of the extremities will fall long before that of deep body tissues, and the unprotected skin of an appendage can rapidly approach ambient temperature. The feet are particularly susceptible if they are inactive and dependent. In such circumstances the muscle pump, which plays such an important role in squeezing the venous blood flow "uphill" to the heart, is almost nonexistent. Backpressure thus occurs on the venous side of the nutritional capillaries, producing a relative stagnation in tissue blood flow. Local swelling of tissues caused by a combination of poor circulation and dependency worsens this condition. Thus a vicious circle is established. In starving survivors, a deficiency of circulating protein contributes further to tissue swelling in the feet, because of the reduced osmosis less fluid moves back into the circulation from the tissues. Fear and anxiety are also predisposing factors for cold injury, as is anything else that decreases peripheral blood flow (for example, footwear or boots that become tight because of swelling of the dependent feet).

The sequence of events in peripheral cooling is as follows: an initial decline in function as tissue temperature falls, numbness progressing through to a total loss of function, injury, and possibly cell death. Injuries produced in such a manner are termed *cold injuries*. They are further subclassified as *freezing cold injuries* (FCI), synonymous with frostbite, and *non- freezing cold injuries* (NFCI), synonymous with *immersion foot*

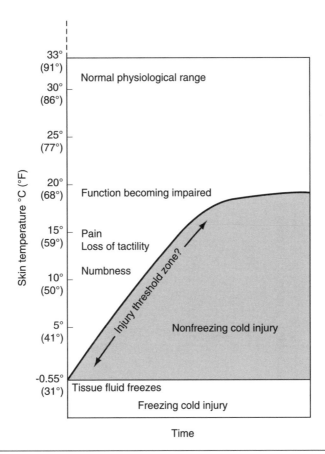

Figure 10.1 Diagrammatic representation of the relationship between cell temperatures, duration of exposure, and pathophysiological consequence.

or, if incurred on land, *trench foot*, depending on whether freezing of tissue fluid has occurred (figure 10.1).

Wind Chill

Wind chill is a major contributory factor in the etiology of cold injury. Relative air movement disturbs the boundary layer of air (forced convection) around the body and increases heat loss. This is the basis of the commonly used term *wind chill*. Siple devised the wind chill index (WCI, table 10.1) to show the physical relationship between temperature and wind speed. People embraced the concept because it corresponded well with their subjective experience in the cold. But the index lacks a scientific basis in physical and biological terms because the cooling rate of a clothed living body in a cold environment is not entirely comparable

Table 10.1   Wind Chill Chart

Beaufort scale	Wind speed knots (mph)	Actual ambient temperatures measured °C (°F)							
		5 (41)	0 (32)	−5 (23)	−10 (14)	−15 (5)	−20 (−4)	−25 (−13)	−30 (−22)
Calm, light air	<3.5 (<4)	5 (41)	0 (32)	−5 (23)	−10 (14)	−15 (5)	−20 (−4)	−25 (−13)	−30 (−22)
Light breeze	5 (4–7)	3 (37)	0.5 (23)	−7 (19)	−12 (10)	−17 (1)	−23 (−9)	−29 (−20)	−33 (−28)
Gentle breeze	10 (8–12)	−2 (28)	−9 (16)	−14 (7)	−20 (−4)	−26 (−15)	−32 (−25)	−38 (−37)	−45 (−49)
Moderate breeze	15 (13–18)	−5 (23)	−11 (12)	−18 (0)	−23 (−10)	−32 (−25)	−39 (−38)	−45 (−49)	−51 (−60)
Fresh breeze	20 (19–24)	−8 (17)	−16 (3)	−23 (−9)	−28 (−19)	−37 (−34)	−44 (−48)	−49 (−57)	−57 (−70)
Strong breeze	28 (25–31)	−10 (14)	−18 (−0.5)	−26 (−16)	−31 (−24)	−40 (−40)	−47 (−53)	−56 (−68)	−62 (−80)
Half gale	35 (32–38)	−12 (11)	−20 (−4)	−27 (−17)	−34 (−29)	−42 (−43)	−49 (−57)	−57 (−70)	−64 (−83)
Gale	43 (39–46)	−13 (9)	−21 (−6)	−30 (−22)	−35 (−30)	−43 (−45)	−50 (−58)	−58 (−72)	−65 (−85)

with the cans of water Siple used in his experiments, a point that he himself recognized. Thus, the application of the WCI to the clothed body unnecessarily exaggerates the danger. It is worth remembering that wind chill slows the cooling power of the environment, not air temperature. It is impossible for skin temperature to fall below ambient temperature, regardless of wind speed.

Although ambient temperature must be well below freezing to produce an FCI of naked skin, the same is not true for NFCI. Tissue cooling to NFCI-threshold levels may occur at higher ambient temperatures if the insulative value of the clothing decreases through compression from wind or wetting. Evaporative heat loss, which is enhanced by forced convection, will further extract heat from the surface of the clothing, thereby increasing the thermal gradient across it.

Non-Freezing Cold Injury (NFCI)

Tissue temperatures between about 17 degrees Celsius (63 degrees Fahrenheit) and –0.55 degrees Celsius (31 degrees Fahrenheit) lasting for a protracted period can result in a non-freezing cold injury. For those whose feet are in water, the colder the water, the higher the risk of injury (Ungley et al. 1945). But NFCI can occur without immersion of the feet. They do not even have to be wet, although evaporation of water from the surface of the foot will greatly enhance cooling and increase the likelihood of sustaining a cold injury in adverse conditions.

The Effects of NCFIs

In the habituation evaluation trials of the British naval life raft in 1976, 15 subjects spent 72 hours in three enclosed 25-man life rafts (5 in each) on water at 5.5 degrees Celsius (42 degrees Fahrenheit) with the mean outside air temperature at 4.5 degrees Celsius (40 degrees Fahrenheit). The mean nocturnal air temperature inside the closed-down rafts was 11 degrees Celsius (52 degrees Fahrenheit). The subjects, living on survival rations of 100 grams of barley sugar and 568 milliliters (1.2 pints) of water a day, remained dry throughout and periodically massaged their feet through their thick wool socks. Within an hour of coming ashore and after a hot shower, 10 of the 15 subjects complained of "hot flushing and tingling of the feet." All were symptom free within 24 hours, and the feet were unremarkable in appearance. However, approximately eight days later, 10 complained of pain and tenderness of the feet, especially in bed at night and on weight bearing first

(continued)

(continued)

thing in the morning. Some reported dulled sensation in the tips of the toes. In all but 3, the symptoms persisted for four to five days before gradually declining. In the remaining 3, some swelling of the feet occurred, and the pain on weight bearing was more severe, necessitating bed rest for 2 of the subjects for about a week. Subsequently, all 10 made an uneventful recovery.

The tissue temperatures in the feet of these 10 subjects clearly fell below the injury threshold and produced a moderate NFCI. Undoubtedly, the state of nutrition and hydration of the subjects contributed to the injuries. In cold conditions, particularly when intake of food and fluid is restricted, metabolic heat production will fall and circulating blood volume will decrease. As we saw in chapter 8, the body in starvation is less able to defend deep body temperature through shivering.

The precise injury threshold for NFCI, both in terms of the absolute tissue temperature and duration of exposure, is unknown. In any case it varies with the factors already noted above. Even in relatively warm waters (20 to 22 degrees Celsius; 68 to 72 degrees Fahrenheit), survivors whose feet have been dependent and partially immersed for several days often have discolored and swollen feet when rescued. These latter injuries probably relate more to poor circulation than subnormal temperature, although it is possible that temperature plays a contributory role.

In a typical case of NFCI, the exposed limb (or injured part) will feel uncomfortable, or even painful, before going numb and suffering impaired function. If the limb has been immersed, the skin will be swollen with a classic wrinkled appearance. It is usually impossible to feel pulses in the affected limb, and sensation will be blunted or absent. Nevertheless, the victim may sense pain when bearing weight. On rescue, if the water temperature was below about 12 degrees Celsius (54 degrees Fahrenheit), the skin will be bright red *(cold vasodilatation)* with occasional purple blotches. A pressure point made by a finger will result in a corresponding blanched area. This area will take some time to redden as blood slowly refills the small blood vessels emptied by the pressure. In contrast, in hot, red skin *(heat vasodilatation),* such a blanched pressure area will refill immediately as circulation rapidly returns to the emptied capillaries. In water above 12 degrees Celsius (54 degrees Fahrenheit) after less extreme cooling, the skin will be waxy white or, if the injury has been present for some time, mottled blue or even blackened.

As the survivor rewarms, circulation will usually return—except in the severest cases—and will be accompanied by a tingling sensation or "pins and needles" *(paresthesia)* and pain. This condition will vary in intensity depending on the severity of the injury. In many cases, strong painkillers will be necessary to alleviate the pain of rewarming. At this stage the pulse will return, accompanied by a hot, throbbing sensation. The limb may become swollen. Some sensation will also begin to return, except possibly to the tips of the digits. Gradually the pain will fade, but a residual increased sensitivity to cold will remain. This condition can be debilitating and may affect future employability and personal circumstances (clothing requirements, ambient heating, interpersonal relationships, etc.). In severe cases, the major blood vessels supplying distal parts of the limbs may be blocked, resulting in gangrene and possibly necessitating amputation. In most cases, some residual damage to nerves and circulation remains as chronic after-effects that can last for life.

A mild form of the injury occurs quite commonly in people indulging in recreational activities in cold environments. The injury produces symptoms akin to those described earlier in the life-raft trial subjects. In such cases the appearance of the skin is often unremarkable.

Therefore, NFCIs can result from exposure to cold ambient conditions above or below freezing. For freezing cold injuries (FCI) to occur, the ambient temperatures *must* always be below freezing and are usually considerably lower. But FCI can occur in exposed skin at temperatures just below freezing if wind chill is present.

Freezing Cold Injury (FCI)

After exposure to extreme cold (below freezing), particularly when the wind chill index is high, the temperature of exposed peripheral tissues may drop below –0.55 degrees Celsius (31 degrees Fahrenheit), the point at which tissue fluid freezes, although skin can freeze at –0.53 degrees Celsius (31 degrees Fahrenheit).

Individuals immersed in seawater near freezing face a theoretical danger of suffering frostbite because the water temperature will be below the freezing temperature of tissue fluid. Seawater does not freeze until around –1.9 degrees Celsius (28.6 degrees Fahrenheit), although this varies with salinity. But reports of freezing cold injury in survivors immersed in water at these temperatures are rare, perhaps because the victims do not survive long enough to suffer the effects of frostbite.

Howard Blackburn's Battle With Frostbite

Perhaps the most notable account of frostbite in a sea survivor is that of Howard Blackburn whose heroic efforts have since entered the realms of folklore.

(continued)

(continued)

On 25 January 1853, Howard Blackburn and Tom Welch were engaged in cod fishing from a dory on the Grand Banks when they became separated from their mother ship, the schooner *Grace Fears*. In the freezing cold they decided their best hope of survival was to attempt to row the 322 or so kilometers (200 miles) to Newfoundland. By the following day, after Blackburn had lost his gloves while bailing, he realized his hands were beginning to freeze and would soon prevent him from helping with the rowing unless he acted quickly. Altruistically, he got hold of the oars and fixed them firmly in his grip until his hands froze into that conformation. Thereafter, he was free to help with the bailing as well as being able to row, using the clawlike configuration of his frozen hands. Welch died from cold on the night of the 27th, leaving Blackburn to continue on his own with the rowing and bailing. After six days without eating or drinking he reached safety. Although he survived and lived for many years thereafter, he lost all his fingers and toes from frostbite.

Howard Blackburn's account illustrates that it is possible for survivors at sea to suffer frostbite, but its occurrence appears to be rare. For modern-day survivors the chances are probably even less likely, although a combination of subfreezing temperatures blown off an ice shelf coupled with high wind speeds makes it possible. Therefore, frostbite remains a threat to fishers in Arctic waters. It is also a threat to Southern Ocean sailors, who should take precautions against possible cold injury if they intend taking an extreme southerly route.

Survival in Arctic Waters

One remarkable account of survival in Arctic waters during World War II was reported by MacDonald Critchley (1943), although the report makes no reference to any freezing injury.

Case 243: A corpulent man of 29, clad in indoor clothing and heavy overcoat, remained swimming in the Barents Sea in July for 9 to 14 hours after his ship was sunk. Two others, who abandoned ship with him, died after 2½ hours. The sea temperature was −1.5 degrees Celsius (29 degrees Fahrenheit)!

Treatment of Cold Injuries

The old adage "Prevention is the best cure" is particularly applicable to cold injuries. There is currently no known cure for cold injury. Even amelioration of its troublesome symptoms can be difficult. Consequently, management of cold injuries is directed primarily at prevention and, after rescue, reversal of the continued cooling to halt progression of the injury.

Prevention

Preventive measures include understanding the mechanisms that cause the injury and the steps you can take to maintain tissue temperatures above the theoretical injury threshold. We discussed these subjects in the previous chapter. Besides keeping the limbs as warm as possible, it is important to loosen bootlaces and remove any constriction around the limbs that may impede circulation and, if possible, elevate the feet. Many advise foot massage, but in the absence of adequate circulation in the upper segments of the limbs, it is difficult to understand how this can be of any significant value.

Treatment

After the rescue of survivors with cold injuries, the possibility that they may also be suffering from hypothermia should not be overlooked. Because profound hypothermia may be life threatening, it follows that its treatment must take precedence over local cold injuries.

The initial first-aid treatment of survivors with FCI involves the removal of clothing and whole-body rewarming in agitated warm water (at a maximum of 40 degrees Celsius; 104 degrees Fahrenheit, i.e., elbow warm but not too hot) until the injured part is completely thawed. This can take a long time. If clothing is frozen to the skin, it should be gently removed as it thaws. If whole-body rewarming is not feasible, the frozen limb should be immersed in a bucket of stirred, warm water (40 degrees Celsius; 104 degrees Fahrenheit). A dilute antiseptic solution should be added to the water, especially if the skin is damaged. Strong painkillers should be available to alleviate the pain associated with the returning circulation. Once thawed, the injured limb should be removed from the warm water and treated in the way that NFCI is treated, described on page 230. During thawing, blood and plasma will leak through the walls of the local small blood vessels that are invariably damaged during the freezing process. This will result in blood-filled blisters, which will eventually shrink and dry, leaving hardened, blackened skin. This skin is sometimes misdiagnosed as a sign of gangrene. If left in place, it will eventually slough, leaving behind underlying viable, although injured, tissue.

Non-freezing cold injuries or freezing cold injuries that have thawed are usually cold and swollen and *must not* be rapidly rewarmed. The swollen limb should be elevated to facilitate local drainage of tissue fluid and insulated to prevent continued cooling. The remainder of the body may be rewarmed to facilitate early vasodilatation of the limb blood vessels, which will facilitate rewarming of the injured peripheral tissues. Those treating the injury should protect the skin from damage and treat any broken skin or wound for infection. Therefore, they should not rub or massage the affected area.

Once rewarming is complete, the area should be dried and exposed to warm air. The victim should avoid bearing weight and be placed in a warmed bed, with the limbs slightly elevated to just above heart level to reduce existing swelling and prevent further swelling. A cage or some suitable device will be required to remove the weight of the bedclothes from the injured part and facilitate the circulation of air to the injured area. A wedge made from a small portion of a clean dressing and placed between the roots of the digits will facilitate circulation of air around the skin of the toes. Provide fluids and nutritional support along the lines outlined in chapter 11. Medical advice should be sought about specific, longer-term treatment.

HEAT ILLNESSES

Heat illness is the term given to describe the feeling of general malaise associated with a rise in deep body temperature from other than pathological causes such as infections. Heat illness occurs when heat gain by the body exceeds heat loss, despite the best efforts of the thermoregulatory system. Consequently, body temperature rises and will continue to do so until some alleviating measure reestablishes heat balance. The very young or elderly are particularly prone to heat illness and need to take special care in hot environments. Regardless of age, anyone already suffering from an underlying infection, which may have produced a preexisting rise in body temperature, is at special risk in hot environments.

As we saw in chapter 2, heat loss from the skin by the evaporation of sweat is essential for the control of body temperature in hot conditions. Thus, anything that interferes with either the production or evaporation of sweat will hasten the onset of heat illness. Dehydration can affect the production of sweat, while the microclimate (e.g., humidity) can affect its evaporation (chapter 3). With the onset of heat illness, both physical and mental performance are significantly impaired, with even relatively easy tasks proving very difficult. Sufferers may be unaware of the severity of their condition, particularly when their temperatures are rising quickly and they are concentrating on a task. Such individuals cannot, therefore, be relied upon to identify that they are developing, or have, a problem.

The early stage of heat illness is often termed *heat exhaustion*. If it is untreated, heat exhaustion can, in the right conditions, progress to *heat stroke*, which is rapidly fatal. Continued exposure to less intense heat, insufficient to cause heat stroke, may however result in chronic heat exhaustion (table 10.2). In the context of survival at sea, those who exercise while wearing special protective clothing are at particular risk of heat illness. This group could include those fighting a fire or conducting other damage-control measures. Dehydrated resting occupants of a survival craft in a hot, humid environment are at risk of heat exhaustion but are unlikely to develop heat stroke.

Signs and Symptoms of Heat Illness

The signs and symptoms associated with heat illness are shown in table 10.2.

Heat Stroke

The most serious form of heat illness is heat stroke. It requires immediate medical attention. Heat stroke is essentially a failure of the body's

Table 10.2   Signs and Symptoms of Heat Illness

Heat exhaustion	Chronic effects (of heat)
Light-headedness, dizziness, faintness	Lassitude
Rapid, shallow breathing (hyperventilation)	Discomfort
Rapid, thready pulse	Irritability
Pins and needles of fingertips and around mouth	Appetite suppression
	Impaired physical and mental performance
Feeling very hot (very uncomfortable)	
	Muscle cramps
Hot, flushed skin, sweating	
	Heat stroke—usually, but not necessarily,
Nausea and vomiting	characterized by hot, dry skin quickly
	leading to unconsciousness, brain damage,
Visual disturbances	and death
Headache	

heat-regulating mechanism, usually caused by overloading the system, ignoring the early symptoms of dehydration and heat exhaustion, and not taking the appropriate corrective measures. Diarrheal illnesses are often a precursor to heat stroke, especially when the necessary ambient conditions are also present.

In heat stroke, sweating usually ceases, resulting in hot, dry skin. Body temperature therefore rises unchecked, leading to hyperthermia and death fairly quickly at a deep body temperature around 44 to 45 degrees Celsius (111 to 113 degrees Fahrenheit). This medical emergency will not wait for treatment ashore. In rapid-onset heat stroke involving hard exercise, sweat may still be present on the skin. In these cases, the absence of the "classic" symptom of hot, dry skin does not mean that the victim is not suffering from heat stroke. Therefore, it is wise to assume that all heat casualties with some impairment of consciousness are potential heat-stroke victims until proven otherwise.

In a survival situation, limited measures are available for the treatment of heat illnesses but, for the sake of completeness, we include here a description of the treatment.

Treatment of Heat Casualties

The best treatment is prevention. Awareness of the potential danger and adherence to proper hydration policy will not only reduce the threat that heat illness will develop but also improve performance in the heat. Thus, people working in conditions commonly associated with heat stress should operate on a work-rest rotation that enables them to off-load body heat before they reach illness levels.

Treatment: Prevention of Dehydration

Dehydration can reduce skin blood flow, which impairs heat transfer from the heat-producing tissues to the surface of the body. Dehydration may occur because of severe water loss; a person can lose up to two liters (about two quarts) of fluid per hour through sweating. Thirst is a poor indicator of dehydration. Dehydration can be well established before a person feels thirsty. Likewise, when rehydrating, the sensation of thirst disappears before the body fully rehydrates. Alcohol increases dehydration by removing more fluid from the body through the kidneys than is consumed with the alcohol in the first place. Thus the normal inefficiency of people with hangovers intensifies if they are working in a hot environment. They are also more prone to heat illnesses than those who are normally hydrated.

A loss of around 1 percent of body weight through dehydration is associated with a significant increase in deep body temperature during exercise, when compared with exercise at a similar workload when fully hydrated. Dehydration also retards the rate of recovery from heat exposure. The adage "water poured down the throat is much more effective at controlling deep body temperature than water poured over the head" is applicable to such situations. To prevent dehydration from occurring in the first place, it is important to keep in fluid balance and ahead of the fluid being lost through sweating. Dark urine means that the body is dehydrated; the aim is to keep the color of urine pale. In normal circumstances, a rehydrating drink should ideally contain sodium and sugar in the proportions shown in table 10.3. This drink is *not* suitable as a survival ration because it includes salt (see chapter 8).

The heat load on the body can come from the environment or through metabolic activity. Consequently, those exercising in a hot, humid environment or those working in a cooler environment while wearing protective clothing are at particular risk of suffering heat illness. In both situations, the body is unable to off-load heat through evaporation, but sweating continues as body temperature rises. This "useless" (unevaporated) water loss intensifies dehydration. In this situation, correcting the dehydration may help maintain circulating blood volume but will do little to control the rise in deep body temperature because the body is unable to evaporate sweat. Continuing to work under such circumstances can produce heat illness even in the Arctic.

Table 10.3   Recommended Makeup of Oral Rehydrating Fluid

Half a teaspoon of salt

Five level teaspoons of sugar

Mix with a liter (quart) of water.

For taste, add a small amount of fruit squash if desired, but the drink should not be too sweet.

Active Treatment

People experiencing the early signs of heat illness must stop work immediately and remove themselves from the heat source. They should undress down to their underwear and rest, preferably lying down, in the shade. If possible, they should expose the naked skin to a breeze to assist in the evaporation of sweat. Fluid (table 10.3) should be freely available to them, and they should drink until they are rehydrated. When the person recovers and passes normal-colored urine, he or she must be cautious about reexposure to heat on the same day. Heat casualties are prone to recurrence in the short term.

If the casualty has reached the stage of collapse and has hot, dry skin, it is imperative to commence an artificial sweating procedure. The person providing aid should lay the victim down, raise the legs slightly, and remove the outer clothing down to the underwear. The underwear, as well as any exposed skin, should be soaked with cool (but not icy) water. It is better to err on the warm side of cool than risk having the water too cold (avoid vasoconstriction of the skin). Again, exposure to a breeze (e.g., under an awning rigged on the upper deck, preferably lying on a hammock made from netting) will accelerate the rate of evaporation. Using the increased heat-exchange capacity of the scalp will optimize body heat loss. These procedures must be maintained until consciousness returns. Because such patients are invariably severely dehydrated, it is important to get fluid (table 10.3) into them as soon as they can safely swallow. Many heat-stroke victims develop secondary infections from their bowel; therefore, a broad-spectrum antibiotic should be given at an early stage, if available. Heat-stroke patients need urgent transfer to a hospital for intravenous fluid replacement and advanced care. Helicopter evacuation should be requested if possible.

Sunburn

Survivors in open boats are at particular risk of severe burning from direct and indirect solar ultraviolet (UV) rays. This UV radiation is partially absorbed and scattered by the earth's atmosphere before reaching sea level. The distance and obliqueness of the pathway it takes through the atmosphere will determine the dosage received. Thus, the dose received in equatorial waters is significantly greater than that received at higher latitudes.

Many people associate the UV threat with the presence of direct sunlight and infrared heat. A significant UV threat may be present, however, in the absence of bright light or strong infrared waves. Severe sunburn can occur on cloudy days or in the presence of strong cooling air movement.

About 50 percent of the UV exposure at the earth's surface is due to indirect, or reflected, radiation caused by the scattering of UV waves. On land, buildings, mountains, or trees provide some protection from harmful UV rays. At sea, however, exposure to a 360-degree arc of direct and indirect UV waves is potentially dangerous unless people take adequate precautions. Many mistakenly believe that taking shelter under an awning will provide protection. Although such a measure will protect from infrared and direct UV rays, it will do little to protect from indirect UV rays reflected from the surface of the water. The maximum reflected dose from this route occurs at about midday and can amount to between 5 and 10 percent of the total unprotected dose at that time.

Thus, skin protection in the form of clothing and high-factor blocking creams is necessary. Because UV waves are able to penetrate loosely woven fabrics, blocking cream should not be confined to unclothed areas. Even when swimming, sun-blocking creams are necessary for protection, especially in those with fair skin who are poorly tanned. Those in a life raft should shelter from UV waves by optimizing the shade provided by the canopy.

In the absence of appropriate sunglasses, UV waves may inflame and damage the eyes. To exclude reflected light, sunglasses may need to have side shades. In tropical regions, those without sunglasses should avoid looking at the reflections on the sea. Instead, they should look through partially closed eyelids or use some form of filter, such as an opaque or loosely woven fabric (e.g., a bandage).

If despite such measures the eyes become inflamed, painful, feel as if they contain sand, or are sticky with pus, the victim should rinse them several times a day with small amounts of fresh water and cover with a bandage to exclude light for at least two days. Thereafter extra care will be necessary to prevent a recurrence.

PSYCHOLOGICAL CONSIDERATIONS

Many anecdotal accounts of epic survival voyages describe, at some length, behavioral (psychological) changes among survivors. Some acknowledgment of psychological considerations is therefore appropriate in a book on the medical aspects of survival. But the very existence of a separate section titled "Psychological Considerations" is misleading; it gives the impression that survivors can deal with such issues in isolation from other factors, such as medical and physiological state, knowledge, experience, and training. Although many authors have attempted to make such a distinction, it is much more instructive to consider the psychology of survival in the wider context of the psychophysiology of survival.

In a survival scenario, the boundary between psychological and physiological responses becomes blurred because many of the signs and symptoms associated with both are similar and, therefore, difficult to distinguish. We know that the physiological state can alter perception. For example, hypoglycemia (low blood sugar) and cold adaptation improve thermal comfort in a cold environment—dangerous alterations if they occur in the face of rapidly falling deep body temperature. Hypothermia will usually produce introversion; dehydration and hunger cause lassitude; and hyperventilation is associated with panic. Furthermore, training and experience can influence the psychological response observed in a survival situation.

The Value of Training and Experience

Nicholl (1960) cites the following interesting example of the value of previous knowledge and training.

During survival tests by the U.S. Navy at Argentia during World War II, one subject, unbriefed on what to expect and do in water at 5 degrees Celsius (41 degrees Fahrenheit), cried in genuine agony after $3\frac{1}{2}$ minutes and had to be hauled out. A second subject, who had been briefed, endured the same temperature for 40 minutes, swimming gently until he almost lost consciousness. Subsequent records of shot-down pilots picked up after relatively long immersion in the winter seas during Korean War operations bore out the value of careful training in these matters.

Historically, numerous accounts tell of communities or groups of individuals who have endured long periods of hardship associated with thirst and hunger. Such accounts demonstrate that maintenance of good morale is often the best ally in times of crisis. Without it, many have perished long before either food or water has run out or before they have reached their limits of physical endurance. An ancient seaman's guide to survival advises that "Hope, perseverance and subordination should form the seaman's great Creed and Duty as they tend to banish despair, encourage confidence and secure preservation."

The term *will to survive* has been much used over the years, but it is of little practical use unless one knows how to acquire it in amounts greater than normal and how to retain it in the face of adversity. Relying on the will to survive can also be dangerous if one regards it as a substitute for good planning and preparation, good survival equipment, and carefully programmed training. Danger also awaits those who feel they have good survival instincts and consequently adopt an attitude of infallibility or immortality. Good leadership is another essential ingredient for successful survival and, as many accounts reveal, the best leaders are often not those who were in command at the outset of an incident. Many of these issues are discussed in an excellent book by Bennet (1983) on the psychology of behavior and survival in arduous conditions, when individuals are pressed to their limits.

For a contemporary view of the psychology of survival, the reader should refer to the work of Dr. John Leach (1995). His model of behavior in a disaster has four phases:

1. Period of pre-impact
2. Period of impact

3. Period of recoil

4. Period of post-trauma (see chapter 11)

Pre-Impact

During the period of pre-impact, people perceive threats and raise warnings and alarms. The response of many individuals during this phase is denial. This period is often associated with inactivity, because people either assume that a real disaster is not about to unfold or fail to grasp the gravity of the situation. In either case, precious time is lost during which the disaster could be averted or important survival actions taken. Such behavior was evident among the *Titanic* crew and passengers and, more recently, among those on *Estonia*.

Initial Disbelief

"A number of people reacted incredulously to the very early signs. They slowly realized that the sounds they heard were abnormal, or rather, they failed to persuade themselves that the situation was still normal" (MV *Estonia*, Final Report 1997).

Impact

During the period of impact, disaster actually strikes and the victim realizes that a real threat exists. The behavior of people during this often-short phase falls into three broad patterns. Some 10 to 20 percent of individuals remain calm and aware and respond in an appropriate and effective way.

Remaining Calm

"A majority of those rescued . . . seem to have grasped the seriousness of the situation when the blows [banging of bow doors] and the list came. They also promptly understood what to do and thus reacted clearly and appropriately. Not without fear, they yet managed to remain rational and to move effectively" (MV *Estonia*, Final Report 1997).

About 75 percent become stunned and bewildered. They have impaired reasoning and restricted attention. They behave in an automatic or mechanical manner, and their sense of the passage of time is affected. Physiological responses include hyperventilation, rapid heart rate, and weakness. As described in chapter 4, these responses also occur on initial immersion in cold water and are part of the fight-or-flight response. Sweating, nausea, and trembling also occur in this period. Finally, 10 to 15 percent of individuals behave in a highly inappropriate way during this period. Their actions are ineffective and can increase the risk to their lives and others. Their responses include weeping, confusion, and paralyzing anxiety. In extreme cases vomiting, urination, and defecation can occur. Panic rarely occurs, except when individuals perceive themselves as trapped.

When Panic Sets In

"The large number of people and their various reaction patterns also created an obstacle to evacuation. . . . Many were seen just holding on without moving; yet others appeared paralysed and seemingly unable to understand what was happening. From the very start of the list many were reported to be passive and stiff, despite reasonable possibilities for escaping. . . . A great number of people were panicking, i.e., behaving without control, and screaming. Some of these were moving but not in a rational or purposeful way. . . . A number of people were shocked and seemingly unable to understand what was going on or what to do. Some of these seem to have been incapable of rational thought or behaviour because of their fear, and screamed or moaned helplessly; others appeared petrified and could not be forced to move" (MV *Estonia,* Final Report 1997).

The prospects for survival are clearly significantly enhanced if a survivor is one of the 10 to 20 percent of individuals who react calmly, appropriately, and effectively in an emergency. The important practical point is that these individuals usually have had training or have had successful outcomes in similar situations. The lesson is clear. Consideration of and training in the appropriate actions to take in an emergency can improve performance at the critical time. An important part of this training is acquiring an understanding of what is likely to happen in an emergency and how to cope. Training should include the following:

▸ Awareness of emergency signals and familiarity with escape routes.
▸ Knowledge of how survival equipment is likely to function, or malfunction, and what is important and unimportant. For example, knowledge that almost all dry suits leak a little in a survival situation and that this leakage is of little significance may prevent the

depression and panic that comes with the belief that one has the only leaky suit (chapter 7).

▶ How to cope with not being seen by a passing ship.

▶ How to use knowledge of survival physiology to overcome the temptation to drink seawater.

▶ How to react to treating family, friends, or other survivors and how to cope with their death.

Recoil

During the period of recoil the immediate physical threat has subsided but secondary threats may exist. For example, survivors may have escaped from the sinking ship but now find themselves in the water or a life raft. The recoil period typically begins with confusion, followed by a slow realization of what has occurred and the damage associated with it. Reasoning ability and recall gradually return. Emotional expression occurs, often in the form of fear, resentment, anger, and guilt. Amnesia is not uncommon. Again, the most common reaction is denial. Apathy frequently follows. The individual responses demonstrated include irrational behavior and psychological disintegration, characterized by a low level of activity, depression, cognitive shutdown, apathy, and resignation to death. A common observation during World War II was "those that died seemed to make no effort" and "they just lay at the bottom of the raft and gave up hope from the start" (Critchley 1943). Physical exhaustion can contribute to this behavior.

The Struggle With Apathy

"Others were apathetic and some only held on to something without making further efforts to save themselves. . . . Some panicking, apathetic and shocked people were beyond reach and did not react when other passengers tried to guide them, not even when they used force or shouted at them. . . . Others seemed to have ceased struggling at some stage, as if giving themselves up for lost. Some have stated that they also, at some time, felt a strong urge to give up although they still possessed some strength. This strong feeling came over them when they suddenly felt their situation was hopeless. Overwhelmed, they lost all mental and physical strength and became passive. They regained their strength and willpower after coming to think of their loved ones, especially their children" (MV *Estonia,* Final Report 1997).

(continued)

(continued)

Other accounts report on events that occurred in the life rafts.

"Somewhat later another man started to fight with the 'leader' . . . He was hanging on to a rope in the middle of the raft. He suddenly became violent, shouting in English about knives, evidently wanting to cut his way through the canopy."

"The women started to bail with their shoes. They urged others to help but most of the others seemed shocked or apathetic."

"People helped each other but some, wearing underwear only, became quite apathetic" (MV *Estonia*, Final Report 1997).

Two points are worthy of note from these anecdotal accounts. First, the enhancement of the will to survive provided by the thought of loved ones, particularly children, appears to be a common occurrence. This was also a strong motivator for the Robertsons during their epic survival voyage. Second, many of the symptoms ascribed to the phase of recoil (irrational behavior, amnesia, apathy, dulling of consciousness) are also associated with hypothermia. Thus, in situations where recoil occurs with concurrent cold exposure (e.g., in life rafts), the distinction between physiological and psychological causes of the behavior displayed is blurred.

The advice to survivors is to try to stay in control of the situation. The survivor should formulate a simple survival plan. The quality of the plan will depend, of course, on the person's previous knowledge and training. The plan should include manageable, prioritized tasks with achievable goals. The survivor should try to disassociate emotions from reasoning; doing so will allow the person to think and remain calm. The survivor should look first to his or her own survival and then help others. As masterfully demonstrated by Sir Ernest Shackleton, giving survivors something to do as soon as possible is beneficial. Having tasks to perform helps overcome the general feeling of lethargy, takes the mind off what has gone before, and gives survivors a feeling that they have some control of their destiny. The leader should identify the skills and hobbies of other survivors; these may prove useful. A knowledgeable person can prepare fellow survivors for the disappointment that may come when those in a passing ship or aircraft fail to notice distress signals. The leader can warn others of the inevitable urge to drink seawater that may arise in a matter of days and prepare them to summon the mental strength to overcome that urge. A strong leader will also convey a sense of hope to survivors who may otherwise be driven to suicide or simply wallow in despair and die long before they should do so from a physiological cause.

Finally, as will be seen in the next chapter, anecdotal accounts indicate that the condition of survivors often deteriorates after rescue when they hand over their survival to their rescuers. The evidence suggests that the survivor should retain a positive mental attitude until rescue and recovery are complete. Those performing the rescue should encourage this attitude.

Chapter Summary and Recommendations

- An assortment of physical and psychological ailments, either alone or in combination with other stressors, can erode morale or decrease survivability at sea. These ailments include osmotic diarrhea, oil contamination, skin ulcers, thermal (hot and cold) injuries, and seasickness.

- Seasickness is a common problem. It can be demoralizing and energy sapping at a time when survivors need to be both alert and active. Seasickness can also accelerate cooling.

- All survivors should take some form of appropriate medication, preferably before boarding any form of survival craft. Hyoscine hydrobromide remains the drug of choice.

- The extent to which individuals are psychologically prepared for disaster and their psychological response to it can significantly affect their chances of survival.

- In a survival scenario the boundary between psychological and physiological responses becomes blurred.

- The prospects for survival increase significantly if a survivor reacts calmly, appropriately, and effectively in an emergency. Training and experience can influence the psychological response observed in a survival situation.

- The survivor should maintain a positive mental attitude until rescue and recovery are complete.

CHAPTER

11

Search and Rescue
and Treatment of Casualties

"**T**HE FORECAST LOOKED HORRIFYING**—isobars on the weather chart were so close together that they seemed to merge into a solid black line. . . . I had a heavy sense of foreboding. . . . The energy of the storm was incredible—they were hurricane conditions. The hull hummed and shook as we surfed down a wave at 27 knots. I braced myself for the impact at the bottom and we dropped for what seemed an eternity before the crash came. . . . We hurtled into a hole at 45 degrees and stopped dead. The cabin went dark as the sea cut out the light. . . . *Aqua Quorum* had become a yellow submarine. . . . The Satcom bleeped, it was Raphael Dinelli's yacht in trouble at 160 miles to windward in atrocious conditions. . . . But I had to go, I knew that. It was that simple; the decision had been made for me a long time ago by the tradition of the sea. When someone is in trouble you help."

*Peter Goss (1998), in the Southern Ocean
during the 1996–97 Vendee Globe single-handed sailing race*

The tradition of the sea may dictate what you should do in situations such as those faced by Peter Goss, but having the courage to carry it through is another matter. You must also know what to do on reaching the site of the accident. Even in relatively calm conditions, the rescue of survivors, particularly those in the water, can provide a stern challenge, especially to those who have not considered the problems involved and have not practiced rescue methods. The number of man-overboard (MOB) incidents on large ships may have decreased markedly in recent years through good design and safety practices, but such incidents are still quite common in the sailing fraternity and the fishing industry. As already described in chapter 4, even in moderately cold water it is likely that the immersed victim will become incapacitated relatively quickly and be unable to do much to assist in the rescue. It is important to have a recovery strategy for helping such people. The time to devise such a strategy is not when the victim is alongside the rescuing craft and in imminent danger of being injured by the hull, stabilizer, propeller, or drowned by rebound waves.

Similarly, the recovery of survivors from life rafts is difficult enough when dealing with fit, active people in moderate sea states, but when survivors are in any way incapacitated and incapable of doing much to help themselves, recovery can be extremely problematic. The modern enclosed lifeboats (TEMPSC) carried on ships may be excellent for pre-

Exhausted and cold World War II-era survivors clinging to an upturned boat, unable to help in their own rescue. They were spotted by chance in midocean by a passing warship.

Photograph courtesy of Imperial War Museum. Crown Copyright.

serving the lives of survivors in hostile environments, but they can be a nightmare when it comes to recovering their occupants in adverse seas. It is often wise not to attempt to open the hatch of one of these craft until the weather moderates sufficiently.

The Danger of Opening the Hatch

On 15 February 1980, following the sinking of the U.S. semi-submersible drilling ship *Ocean Ranger* off the east coast of Canada (high winds, severe sea state, water temperature −0.6 degrees Celsius (31 degrees Fahrenheit), air temperature −4 degrees Celsius (25 degrees Fahrenheit), 22 of the 84 crew managed to board a TEMPSC. Unfortunately, when they were subsequently attempting to board the rescue standby ship, the TEMPSC suddenly capsized and flooded. In spite of the heroic efforts of the crew of the standby vessel to save them, none of the 22 crew members survived.

RESCUE COLLAPSE AND DEATH

Apart from the practical difficulties of rescuing survivors, analysis of rescue statistics reveals that in some situations a percentage of those who die as a result of immersion in cold water do so just before, during, or shortly after rescue. The percentage varies between incidents but on average appears to be about 20 percent. Those whose experience is based on the rescue of one or two casualties at a time often dispute the existence of this problem; 20 percent of the one or two is difficult to appreciate as an indication of a larger problem. Thus, only when large numbers are being rescued does the phenomenon become noticeable. For example, in the U.K. immersion-incident survey (Oakley and Pethybridge 1997), 20 percent of those recorded as unconscious at rescue were dead when delivered to medical care. In retrospect one can find evidence of similar incidents in one-off rescues.

The Aymeric

In May 1943 survivors from *Aymeric* found themselves in the icy waters south of Greenland after the ship was torpedoed. An adjacent warship in contact with the enemy submarine called on the rescue ship *Copeland* at

(continued)

(continued)

the tail of the convoy to pick up the survivors. It arrived on scene 40 minutes later, by which time many of those in the water were already dead. Of the 24 survivors rescued alive by *Copeland,* 6 died shortly afterward, and 7 of the 14 rescued by the trawler *Northern Wave* also died.

Even survivors in lifeboats and life rafts, who have not been immersed, may collapse and die during rescue. Many other survivors show a sudden deterioration in their condition without suffering a fatal outcome at about this time. Why does this occur?

At the time of rescue the immersion victim is likely to be suffering from one or more of the following:

▶ Near drowning

▶ Significantly impaired peripheral neuromuscular (nerve and muscle) function

▶ Hypothermia

▶ Trauma

In addition, as we saw in chapter 4, physiological alterations in blood volume and its distribution because of immersion can impair cardiovascular (heart and circulation) function as the survivor is being removed from the water. This event is particularly likely to occur when the water is cold. The mechanism causing these changes is still a matter for debate, but the problem has significant implications for the methods used to rescue and treat immersion casualties. Although some scientists denigrate anecdotal evidence, the many reports from different sources describing survivors losing consciousness and even dying during rescue cannot be ignored. The historical and anecdotal evidence for this problem is outlined below.

Historical and Anecdotal Evidence

Sudden loss of consciousness or death during or following the rescue of immersion victims came into prominence during World War II because of its frequent occurrence. But there are many earlier recorded instances of "circumrescue collapse" and death.

In 1762 the preeminent British naval physician, Sir James Lind, first highlighted the danger of collapse in the post-immersion period. Thirty-six years later James Currie also observed the deterioration of the condition of his experimental subjects post-immersion. Sir John Franklin, in the account of his early expedition to the Northwest Territories in 1821, gives a description of both cold-water swim failure and post-immersion collapse. He described how the expedition's surgeon, Sir John Richardson, became immobilized while attempting to swim across the ice-cold Coppermine River. Because of his incapacitation (swim failure; see chapter 4), his colleagues towed him back to the riverbank. He was conscious and shivering violently on arrival ashore but still sufficiently lucid to advise his rescuers on his treatment (method of rewarming) before he collapsed into unconsciousness. Some time passed before he regained consciousness, fortunately unaffected by his experience.

In 1889 in Germany, a police physician named Reinke, who frequently treated seamen rescued from Hamburg harbor, also recorded the phenomenon of death after rescue.

During the American inquiry into the sinking of *Titanic,* one witness, Fifth Officer H.G. Lowe, described how the crew of his boat rescued four men wearing life jackets from the calm water, one of whom subsequently died. Another witness at the same inquiry but in a different lifeboat, Mrs. E.B. Reynolds, described how two other immersion survivors died in the stern of her lifeboat shortly after rescue from the sea.

Reports of the sinking of the liner *Lusitania* in 1916 contain several accounts of survivors rescued alive and subsequently dying onboard the rescuing fishing boats. In contrast, several "bodies" believed to be dead on rescue and placed under tarpaulins in the bows of some fishing boats subsequently caused alarm to their rescuers when some survivors began to remove the tarpaulin as they regained consciousness! Also in World War I, during the battle of the Falkland Islands, it was reported that most of the 200 survivors of the German battle cruiser *Gneisenau* died onboard one of the rescue ships. In 1974 Ortzen described a survivor from the passenger ship *Vestris,* sunk in 1928, collapsing and dying shortly after climbing into a lifeboat following a period of immersion.

The lessons of World War I apparently went unheeded, and it was not until World War II, with many men being rescued alive from cold water, that the problem of post-rescue collapse and death was again highlighted. For the Allies, Wayburn (1947) recorded some instances of death after rescue among ditched U.S. Air Force personnel. In Germany, Grosse-Brockoff (1946) reported how men rescued alive from Norwegian fjords subsequently died from the "sequelae of hypothermia." Although Grosse-Brockoff gave no figures, it is believed that considerable casualties resulted from this cause in one incident. Also on the German side, the problem of *Rettung Kollaps* ("rescue collapse") became noticeable during the Battle of Britain. The highly efficient German Air Sea Rescue Services recovered a large percentage of the ditched Luftwaffe, British, and American aircrew only to observe many die 20 to 90 minutes after rescue. Investigation of this problem was offered as justification for some of the infamous Nazi experiments (Alexander 1945).

Establishing the precise extent of the problem of post-rescue collapse is difficult because the death certificate often records only the terminal event. Although pulmonary (lung) changes are usually found at post-mortem in drowning cases, the absence of such changes apparently does not influence the postmortem diagnosis. It is estimated that 10 to 20 percent of drownings are "dry drownings" (Moritz 1944). Even immersion cases in which the history suggests that the victim kept the airway clear of the water at all times are sometimes inexplicably classed as drowning deaths! Producing reliable statistics on the contribution of acute cold to these deaths is difficult because the condition leaves no definitive changes at autopsy. Therefore, it may well be that cold or other mechanisms caused some deaths attributed to dry drowning.

One of the few numerical indications of the possible extent of the problem of post-immersion collapse comes from McCance et al. (1956). In their survey of merchant shipping losses during World War II, they found that of a series of 289 shipwreck survivors, 17 percent of the 159 rescued from water at a temperature of 10 degrees Celsius (50 degrees Fahrenheit) or less died within 24 hours of rescue. None of the 129 rescued from water at a temperature above 10 degrees Celsius (50 degrees Fahrenheit) died after rescue. Although the investigators do not give the cause of death of the 17 percent, the association with cold water is significant.

The Problem of Post-Immersion Collapse

MacNalty (1968), in his *History of the Medical Services in War,* suggests that the problem may be even greater and not be exclusive to very cold waters:

"A not uncommon feature observed in other waters as well as in the Arctic, was that many survivors who had managed to get themselves to the point of being helped from the sea collapsed when safety was within reach, and required to be handled in the same manner as those who had been helpless while still in the water."

Keatinge's (1965) account of the sinking of the liner *Lakonia,* off Madeira, when the water temperature was approximately 18 degrees Celsius (64 degrees Fahrenheit), lends support to MacNalty's suggestion. He reports that most of the 15 dead bodies on *Montcalm,* the first rescue ship on the scene, were alive on rescue.

Post-immersion death is not limited to those who are unconscious on rescue. Max Matthes (1946) recounts how ditched aircrew who had been in the sea for relatively short periods and were fully conscious and aiding in their own rescue became unconscious after rescue and, in some cases, died.

The SS Empire Howard

Other classic examples of post-immersion collapse include the survivors of the SS *Empire Howard.* The captain of the rescue ship, Captain Downey, reported that:

"Everyone was conscious when taken out of the water but many lost consciousness when taken into the warmth of the trawler. Nine (out of 12) died shortly after being rescued" (Lee and Lee 1971).

Similarly, nine survivors from HMS *Kite,* sunk while on Arctic convoy duties in 1943, were rescued by a sister ship, *Keppel.* All were conscious when rescued, but only five survived.

Post-war documented examples of post-immersion collapse and death in swimmers are available from several sources, including the records of the British Long-Distance Swimming Association. Hardwick (1962) gives an account of a swimmer's collapse and loss of consciousness after being forcibly removed from the water after 11 hours of swimming in the Irish Sea. Lloyd (1964) describes the sudden collapse and death of a speleologist who shortly before was enjoying a bar of chocolate following rescue in a hypothermic condition from a flooded cave.

Golden (1973) describes two cases of teenage boys dying from ventricular fibrillation in hospital following a boating accident. Although no body temperatures were recorded (their temperatures were below the range of clinical thermometers in use at that time), the circumstantial evidence and the condition of other survivors left little doubt that hypothermia played a major role in the fatal outcome. In another case investigated by one of the authors, a 54-year-old schoolteacher was canoeing on a lake with two young female passengers when the canoe capsized. The three were immersed in water at 12 degrees Celsius (54 degrees Fahrenheit) for about 10 minutes. During this time they held on to the upturned canoe. People in a rowing boat rescued the two girls, but the teacher proved too heavy to be helped aboard and was towed ashore. On reaching shore he complained of cold and numbness of the legs but was otherwise rational and coherent. His wet clothes were removed and he was massaged. A short while later, "he suddenly went rigid and died!"

In June 1979 HMS *Jupiter* was involved in the rescue of survivors from the MV *Iris,* which sunk in the Bay of Biscay in a force 8 gale. The water temperature was 15.5 degrees Celsius (60 degrees Fahrenheit), the air temperature was 12 degrees Celsius (54 degrees Fahrenheit), and the sea state was 6 with an 8-meter (25-foot) swell. With the ship rolling through an arc of 45 degrees, swimmers effected the rescue using a helicopter-type rescue strop attached to an A-frame. Of the 12 people rescued, "one died halfway up the ship's side," and another "died within minutes of rescue."

During the 1979 Fastnet race (water temperature of 15 to 16 degrees Celsius; 59 to 61 degrees Fahrenheit), 3 of the 15 fatalities (20 percent) occurred during rescue—1 while being rescued by helicopter and 2 while trying to climb up a scrambling net thrown over the side of a ship.

Other anecdotal accounts are worthy of note:

▸ The pilot of a light aircraft flying from Sweden to Denmark in winter ditched in the sea within sight of the lights of Copenhagen. The pilot scrambled on to the outside of the cockpit to await rescue. He recalls the rescue helicopter arriving overhead, the rescue crewman being lowered on the winch, and the rescuer placing the strop around him. His next recollection was awakening in a hospital bed in Copenhagen.

▸ A young woman rescued from the sea off the coast of Cornwall, United Kingdom, in summer remembers the helicopter rescue diver putting the rescue strop around her and commencing the lift. She regained conscious-ness sometime later lying in the back of the helicopter.

▸ In 1985 a sailor who was washed overboard from a British navy ship off North Cape (sea state 6; air temperature –2 degrees Celsius [28 de-grees Fahrenheit]; sea temperature 5 degrees Celsius [41 degrees Fahren-heit]) spent eight minutes in the water before being rescued by helicopter using a single-strop lift technique. He was successful on the third attempt in putting the strop around his body. His next recollection was on regain-ing consciousness as he was placed in a horizontal attitude to be maneu-vered through the helicopter door. Subsequently he made an uneventful recovery.

▸ The head of Swedish Maritime SAR Service, U. Hallberg, reported two cases in which death occurred during helicopter winching. In the sec-ond of these, the victim was alive in the water, floating in his life jacket, but after being winched into the helicopter was found to be dead.

▸ As mentioned in chapter 4, a survivor from the Cormorant Alpha helicopter ditching in 1992 was widely quoted in the media when he said, "I think I was about at the end of my rope (sic) when they got me up. I must have been so relieved I lost consciousness." He was winched from the cold (less than 7 degrees Celsius; 45 degrees Fahrenheit) water by a Norwegian helicopter after spending 1 hour and 21 minutes immersed.

▸ In its description of the 1994 *Estonia* sinking, *Time* magazine reported, "Of those who managed to scramble overboard, only 139 survived. The rest died of shock and hypothermia before rescuers could pluck them from the storm-tossed . . . sea; some expired even as they were being winched to safety" (*Time* magazine, 1994).

Firsthand Account of Collapse

At a meeting held at the Swedish Society of Medicine shortly after the *Estonia* incident, one of the helicopter rescue divers provided a classic account to one of the authors of the collapse of a casualty he was attempting to rescue.

"After attaching a single strop to the victim he was lifted vertically from the water. About 2 meters (6.6 feet) from the helicopter the winch jammed, at the same time the victim lost consciousness and began to fall through the strop; the diver held onto the victim for as long as he was physically able, before fatigue prevented him from maintaining his grip any longer, and the victim fell back into the water. On hitting the water the victim appeared to regain consciousness, swim a few strokes and suddenly die despite the efforts of the diver who had reentered the water in an attempt to save the man."

Consideration of the preceding anecdotal evidence suggests that three phases of the rescue process have particular risks:

1. Pre-rescue—just before rescue

2. During rescue—during or immediately following removal from the water

3. Post-rescue—following rescue

Golden et al. (1991) have given the generic title of "circumrescue collapse" to these three stages and have related them to previously used terms. The possible mechanisms responsible for circumrescue collapse and their implications for the treatment of immersion victims are described in table 11.1.

POSSIBLE MECHANISMS INVOLVED IN CIRCUM-RESCUE COLLAPSE

A number of possible mechanisms may be responsible for collapse and death at this time. Some are specific to the phase of rescue, while others are common to all three phases.

Table 11.1   Stages of Circumrescue Collapse

	Stages of rescue		
	Pre	During	Post
Previously used terminology	None	Rescue collapse Post-immersion collapse	Rescue collapse Post-immersion collapse Post-rescue collapse
Other conditions possibly playing a contributory role			Near drowning Rewarming collapse

Phase 1: Pre-Rescue

The reference of MacNalty (1968) to some survivors who at "the point of being helped from the sea collapsed when safety was within reach" implies a sudden physiological deterioration in the condition of some victims just before rescue. Although other anecdotal accounts support the existence of this problem, they are generally difficult to verify. Any hypothesis as to the possible mechanism of pre-rescue collapse is therefore speculative.

The reported sudden onset of the deterioration in the condition of survivors is suggestive of a cardiovascular (heart and circulation) mechanism. The extreme slowing of the heart in hypothermia coupled with an increase in the viscosity (thickening) of cold blood results in a marked reduction in the blood flow in the coronary arteries (the blood vessels that supply the heart muscle with the necessary oxygen and nutrients). McConnell et al. (1977) have shown experimentally that although coronary blood flow declines in profound hypothermia, it is sufficient for the reduced work being undertaken by the slow hypothermic heart. But any increase in the work of the heart (such as anticipatory activity in preparation for rescue) can lead to an insufficiency problem. The reluctance of cold blood to release the oxygen it carries compounds this problem.

An additional problem with coronary blood flow occurs if blood pressure falls. When this happens under normal circumstances, the coronary arteries compensate by dilating to ensure maintenance of an adequate blood supply. In the hypothermic heart this critical safety mechanism is impaired. Consequently, the inner lining of the heart chambers, where the important electrical-conducting circuitry is located, becomes deprived of oxygen. The heart becomes more susceptible to electrical asynchrony and sudden arrest. Catecholamines (blood-borne chemicals released in response to stress), particularly *norepinephrine*, have been shown to have a protective effect in hypothermia. This effect is attributable to the rise in blood pressure that their secretion produces rather than any specific direct effect on the heart.

Ample evidence shows that in humans immersed in cold water, catecholamine secretion greatly increases. It seems plausible that a person who becomes aware of imminent rescue with the arrival on scene of a rescue helicopter or boat may feel an overwhelming sense of relief. This sensation may be associated with a reduction in catecholamine secretions and the withdrawal of their protective effect. In the absence of confirmatory evidence, this hypothesis is purely speculative at this stage, even if it sounds plausible. This mechanism may also be the cause of the sudden deterioration in normothermic victims when they "hand over" their well-being to their rescuers (chapter 10). Thus, this problem may occur at any stage of rescue.

Phase II: During Rescue

Debate continues about the mechanisms responsible for rescue collapse, with thermal, cardiovascular, biochemical, and hormonal responses being suggested as potential causes. The debate is not of purely academic interest. It has significant implications for the methods used to rescue immersion victims and their early treatment. The effectiveness of those procedures can mean the difference between life and death for many immersion victims.

Many studies have examined post-immersion thermal responses and methods of rewarming cooled individuals. In most of these studies, a continued fall in deep body temperature during the initial phase of rewarming, termed the *afterdrop,* has been observed. The magnitude of the afterdrop and the rate of return of deep body temperature to normal have been used as the criteria to evaluate rewarming techniques (figure 11.1).

The preoccupation with the afterdrop has arisen from the belief that it is responsible for many post-immersion deaths. This is in accord with the Nazi researchers at Dachau (Alexander 1945). They believed that the afterdrop was caused by cooled blood returning to the central circulation from the cold periphery during rewarming. They made no effort to differentiate between early and late deaths. The fact that the reported timing of many instances of collapse and death coincided with the nadir of the afterdrop reinforced their hypothesis. The Nazi researchers implied that if a victim was within a couple of degrees of a lethal deep body temperature when removed from the water, the afterdrop could take the person into the lethal zone. Standard texts have widely quoted and

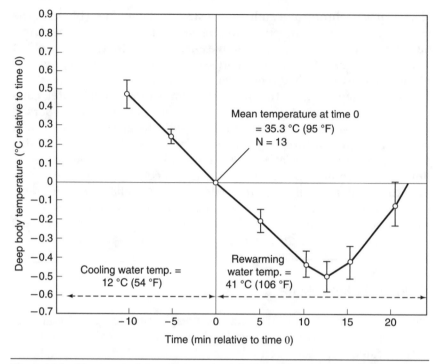

Figure 11.1   The deep body temperatures (rectal) of experimental subjects during cooling following immersion in cold water and rewarming in hot water. The afterdrop in the rewarming phase is evident.

accepted this hypothesis. The hypothesis has also helped form some of the "myths" associated with management protocols (torso-only rewarming) and equipment (rewarming devices) used with hypothermic casualties.

Although this hypothesis may have been credible in the context of the laboratory experiments of the Nazi researchers, it does not fit easily with the anecdotal accounts of the collapse of individuals assisting in their own rescue. To do this, they must have been conscious and active—that is, they must have had a deep body temperature above 32 degrees Celsius (91 degrees Fahrenheit) and more probably above 34 degrees Celsius (93 degrees Fahrenheit). It is not then possible for an afterdrop of 1 to 2 degrees Celsius (2 to 4 degrees Fahrenheit) to lower deep body temperature below 25 degrees Celsius (77 degrees Fahrenheit), the temperature most frequently associated with post-immersion death during the Dachau experiments, or even to 28 degrees Celsius (82 degrees Fahrenheit), the temperature at which cardiac sensitivity to cold enters the danger zone.

The afterdrop hypothesis for rescue collapse went unchallenged for many years, until Golden and Hervey (1977, 1981) demonstrated that it was not necessary to invoke physiological, hemodynamic (related to blood flow) mechanisms to explain it. Instead, they proposed that a continued fall in deep body temperature is an inevitable consequence of the physical laws of conduction of heat in the body when, after being cooled for some time, the body is removed from cold water and placed in hot water. In such a situation, heat will continue to flow from the warmer deep body tissues through the cooler intermediate tissues to the surface, until the thermal gradient reverses (figure 11.2).

This hypothesis is supported by the finding that the afterdrop is greatest when deep body temperature is monitored at sites such as the rectum, where temperature is determined primarily by conductive heat flow, and least when temperature is measured in sites such as the esophagus ("gullet") or heart, where temperature is determined primarily by blood flow. The fact that a significant afterdrop has not been found in the heart itself (Golden and Hervey 1981), even when one is present in the rectum, suggests that during hot-water rewarming, blood returning from the periphery is in fact a source of heating rather than cooling for the deep body tissues. Figure 11.3 shows that 5 and 10 minutes after commencing rewarming, the central venous blood temperature is rising while the rectal temperature is still falling, lending support to the conclusion that the afterdrop is predominantly a phenomenon of the rectal site.

Some scientists still do not accept Golden and Hervey's dismissal of the afterdrop as a cause of circumrescue collapse and death. They offer in support of their argument the continued fall in deep body temperature in experimental subjects on removal from cold environments other than cold water, and when being rewarmed by methods other than immersion in hot water. It is obvious that an "afterdrop" (or more correctly, continued cooling) will occur if rewarming is ineffective. But, despite the tempory continued fall in rectal temperature, cooling does not happen centrally to any significant degree when hot-water rewarming is instituted immediately on removal from the cold water, as in the Dachau experiments and in those conducted by Golden and Hervey (1977, 1981). As such, some of the confusion in the field results from the imprecise use of the term *"afterdrop."* A distinction should be made between the afterdrop in rectal temperature during rapid rewarming in hot water (as originally observed and defined by the Nazi researchers) and that seen with other and/or slower methods (e.g., exercise, spontaneous methods) of rewarming. The term *afterdrop* should not be applied to the cooling observed when victims are rescued and placed in an environment that allows continued cooling (e.g., back of a helicopter).

Golden and Hervey's findings refute the traditional explanation that cold blood returning from the peripheral vascular bed to the central

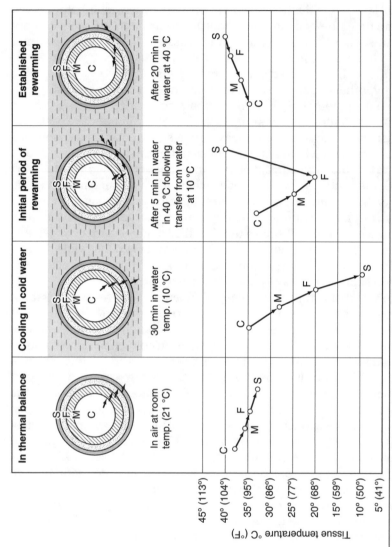

Figure 11.2 Diagrammatic representation of the time course of conductive heat flow through the superficial and deep tissues of the body in air and during immersion in cold (10 degrees Celsius; 50 degrees Fahrenheit) and hot (40 degrees Celsius; 104 degrees Fahrenheit) water. (C = core as measured in the rectal site; M = muscle; F = subcutaneous fat; S = skin; → = direction of heat flow).

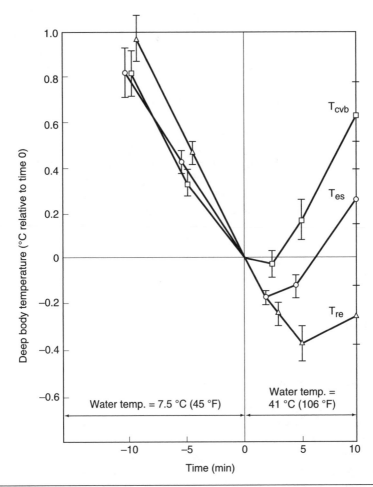

Figure 11.3   The mean (± standard error) temperature recorded in the central venous blood and heart (T_{cvb}), the esophagus (T_{es}), and the rectum (T_{re}) during the last 15 minutes in cold water and the first 10 minutes in hot water. Changes are plotted relative to zero, the temperature at time zero for all three sites (mean = 31.5 degrees Celsius; 89 degrees Fahrenheit); n = 24.

Reprinted from Golden and Hervey (1977).

circulation causes the afterdrop. They also contradict the suggestion that post-immersion death is caused by a "cold bolus" of blood returning to the heart from the periphery during rewarming and causing cardiac problems. Although researchers (Hayward et al. 1984) have reported indirect evidence for this phenomenon in some circumstances, Golden and Hervey (1981) could find no evidence to support it. In any case, it is difficult to see how such a mechanism could account for the transient loss of consciousness reported in some survivors.

Thus, the afterdrop cannot account for the majority of post-immersion deaths because these must occur at deep body temperatures that are well above lethal cardiac levels. In any case, the afterdrop is primarily a phenomenon of rectal rather than cardiac temperature. We conclude therefore that the afterdrop is largely of academic interest and not critically important to either the rescue or treatment of immersion casualties provided the casualty is adequately rewarmed.

An alternative explanation for post-immersion collapse and death therefore seems to be required. Although it is conceivable that a mechanism similar to the one we outlined for collapse just before rescue may also explain collapse during rescue, other factors now come into operation. Identified by the anecdotal evidence, several factors could produce a sudden and dramatic alteration in cardiovascular function:

- ▶ Prolonged exposure (in water or life craft)
- ▶ Mode of rescue (posture of victim, requirement for physical activity)
- ▶ Hypovolemia (decrease in circulating blood volume)
- ▶ Hypothermia

Although listed separately, these factors are inextricably linked and, more importantly, will present a greater hazard when they occur in combination.

Physiological Changes Associated With Head-Out Immersion

The physiological changes associated with immersion in water have already been described in some detail in chapter 4. These changes are primarily the result of a reduction of the influence of gravity, together with the effect of the pressure of the water surrounding the body (hydrostatic pressure). With regard to collapse during rescue, the most important of these changes are those that influence the cardiovascular system and blood volume. In sum, these responses, observed during resting, upright, head-out immersion in thermoneutral water, include

- ▶ a 250-milliliter (8.5-ounce) average enhancement of diastolic filling (filling of the chambers of the heart during cardiac relaxation),
- ▶ an increase in right atrial pressure (pressure of the venous blood on entering the heart) of 12 to 18 millimeters (0.48 to 0.70 inches) of mercury, and
- ▶ a 32 to 66 percent increase in cardiac output (the outflow from the heart).

This increase in cardiac output during immersion is due entirely to enhanced venous return to the heart.

Diuresis (increased excretion of urine) is also invoked by increases in central blood volume following immersion because the body senses *hypervolemia* (increase in circulating blood volume), despite the fact that total blood volume remains constant initially. In fully hydrated subjects, head-out, upright immersion in thermoneutral water can result in urination reaching 350 milliliters (12 ounces) per hour. Immersion in cold water is likely to increase this figure because of cold-induced diuresis and, if severe hypothermia occurs, because of the direct inhibitory effect of cooling on kidney function. These alterations cause the immersion victim to be *hypovolemic* (decrease in circulating blood volume) in absolute terms, although central blood volume adjusts to meet the requirement of the immersed state. As a consequence there is no adverse effect, from this cause, experienced by the individual while immersed. General body cooling also induces a relative hypovolemia through fluid shifts from blood to tissue fluid (chapter 6).

Physiological Effects of Removal From the Water

When removing (rescuing) an individual from water following a prolonged period of immersion, the hydrostatic assistance to circulatory function suddenly vanishes just as the full effect of gravity reimposes itself on the body. The blood volume of the individual may now be inadequate (hypovolemic) for the air environment. Gravity tends to induce a redistribution of blood with venous pooling in the lower limbs when the victim assumes a vertical posture *(postural hypotension)*. The resulting reduction in blood returning to the heart will affect cardiac output and, if not corrected, the person will faint as the blood supply to the brain falls. Under normal circumstances, a reflex *(baroreceptor* reflex) detects and counters falling blood pressure, resulting in constriction of the blood vessels and an increase in heart rate. This reestablishes blood pressure and ensures adequate blood flow to the heart and brain.

Some evidence suggests that moderate cooling impairs the baroreceptor reflex. Thus, in cooled individuals, the compensating regulatory response may be dulled or even absent, permitting a fall in blood pressure to occur. The consequence of this redistribution of blood on removal from the water is compounded by the loss of circulating fluid volume during immersion. The resulting associated fall in pressure may be severe enough to cause a shortage of blood to the brain *(cerebral ischemia)* and temporary loss of consciousness or even a shortage of blood to the coronary vessels of the heart, resulting in cardiac arrest.

Physical effort by the casualty—including swimming, climbing, or boarding rescue craft—is another factor that can greatly increase the demands placed on the compromised circulation at this time. Even in individuals with normal physiological responses, such physical demands will increase muscle blood flow and cardiac work and, as a consequence the requirement for oxygen by the heart muscle. The increased heart rate and blood viscosity caused by shivering and cooling respectively will further increase the demand for cardiac work and reduce coronary artery blood flow.

Based on the evidence presented thus far, an alternative hypothesis to the afterdrop for collapse during rescue involves a sudden deterioration in the condition of an individual because of the collapse of blood pressure. This results from

- ▸ loss of the hydrostatic assistance to venous return,
- ▸ introduction of the full effects of gravity,
- ▸ decreased circulating blood volume,
- ▸ dulled baroreceptor reflex responses,
- ▸ increased blood viscosity,
- ▸ diminished work capacity of the heart,
- ▸ performance of physical work, and
- ▸ decreased psychological stress.

Preexisting coronary disease (narrowing of the coronary arteries) will add another factor that can diminish the supply of oxygen to the heart muscle.

Death probably occurs because of an inadequate supply of blood, and therefore oxygen, to the heart or brain. The consequence of reduced blood flow to the brain may be more variable and can range from periods of dizziness to unconsciousness and death. Another possibility is that some casualties will lose consciousness while climbing nets or ladders and fall back into the water and drown. With regard to the mechanism causing collapse during rescue, it is probably significant that anecdotal accounts report that recovery from periods of unconsciousness occurs once the casualty is horizontal and at rest. Such recoveries suggest that cardiovascular rather than thermal alterations (the afterdrop) are responsible for such collapse.

Testing this hypothesis is difficult because real rescues are diverse and uncontrolled, and the possibility of making the necessary measurements is nil. But Golden et al. (1991) have reported data showing the effects of lifting a subject vertically using a single helicopter strop, after 30 minutes in water at 15 degrees Celsius (59 degrees Fahrenheit). Within seconds of the start of the lift, the central venous pressure of the subject had fallen by 12 millimeters (0.48 inches) of mercury. Blood pressure remained lowered until the subject was reimmersed at the end of the 60-second lift (figure 11.4). The heart rate of the subject increased by 30 beats per minute during the lift. The authors interpreted these changes as the consequence of removing the hydrostatic support provided by the water and reimposing the full effects of gravity.

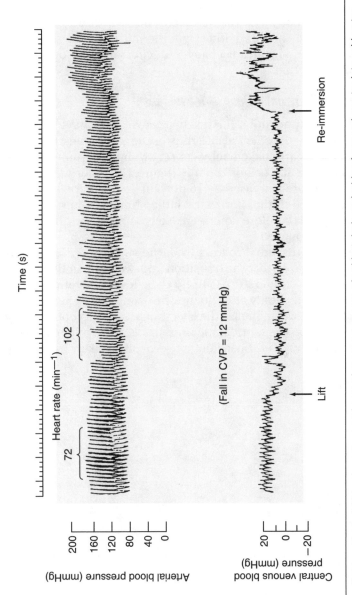

Figure 11.4 Tracing of arterial blood pressure and central venous pressure of subject during winching from and reentry into cold water. Reprinted from Golden et al. (1991).

The immediate increase in heart rate following the fall in central venous pressure seen in figure 11.4 is part of the normal baroreceptor response to a fall in blood pressure. The magnitude of the increase in rate can be regarded as an index of the cardiovascular strain that a rescue technique is placing on the body. On this occasion, the subject did not lose consciousness because he was neither hypothermic nor immersed for sufficient time to have significant changes in circulating blood volume or baroreceptor function.

Vertical Versus Horizontal Rescue Techniques

In view of the implications for rescue, researchers have undertaken studies to examine the effect of differing rescue techniques. Golden et al. (1991) lifted 17 nearly naked subjects (in swimming trunks) from cold water either vertically or horizontally (figure 11.5). During vertical lifting, mean heart rate increased by 16 percent. In comparison only small increases occurred during horizontal lifting. None of the subjects were hypothermic, and therefore none were likely to have their baroreceptor responses compromised.

With regard to rescue, it follows from the suggested mechanism for circumrescue collapse that the transition from water to air during rescue from water at any temperature is likely to be less traumatic in subjects who are lifted horizontally after floating horizontally during immersion. This form of lifting will minimize the effects of postural hypotension (with horizontal flotation the hydrostatic pressure effect is minimal). Vertical lifting of a victim who has been floating vertically is likely to present the

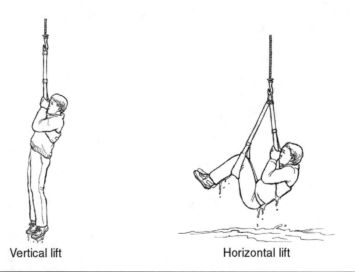

Vertical lift   Horizontal lift

Figure 11.5   Single (vertical) and double (horizontal) helicopter lifts.

greatest hazard because the beneficial effects of hydrostatic pressure are lost just as the deleterious effects of gravity are introduced. This hazard will increase if the water is cold and especially if the victim is hypothermic.

Experience With Horizontal Rescue

"In the latter stages of the war, the German Lifeboat Service noted (albeit serendipitously) that rescuing immersion victims by inserting a ladder beneath them and levering them out in a horizontal attitude, enhanced their survival chances when compared with removing them vertically from the water, or requiring them to assist in their own rescue" (Koffal 1980).

Protective clothing, such as immersion suits, tends to encourage horizontal flotation as well as delay the onset of hypothermia, thereby prolonging survival. Life rafts tend to do the same. During lengthy periods of survival in the cold, victims may become profoundly hypovolemic and, eventually, hypothermic. Thus, although better protected from the effect of hydrostatic pressure than vertically immersed victims, they will be at increased risk of postural hypotension and may also suffer circum-rescue collapse.

Finally, if wave splash is not threatening the survivor's airway, removal from water in the horizontal posture is preferable in all circumstances if rescuers can accomplish it. Any demand for physical effort by the victim at the time of rescue appears to carry a risk of precipitating collapse and death. With the inevitable proviso "circumstances permitting," rescuers should handle immersion victims gently and treat them as the potentially critically ill patients that they are. The authors endorse the efforts that some organizations are now making to modify rescue techniques to meet these principles. Many organizations have found it more practical and safer to rescue a person from the water into a ship's boat in the first instance and then recover both victim and boat back onboard the parent vessel. In that way the victim lies in a horizontal position throughout.

The United Nations International Maritime Organization (IMO) International Convention for the Safety of Life at Sea (SOLAS) regulations require all ro-ro ferries to carry a davit-launched powered boat (fast rescue craft, or FRC) to facilitate the rescue of people from the sea and the mustering and towing of life rafts. Many of these boats are of semirigid inflatable design with a crew of two or three. Because several other organizations use these boats for rescue purposes, we will discuss the problems of rescue and treating immersion victims in the exposed confines of these craft on page 269. For other craft, such as fishing boats and

Wearing a securely anchored safety harness when on the upper deck in bad weather, as the crew of the U.S. Coast Guard cutter *Ponchartrain* is doing, will significantly reduce the risk of falling overboard.

Official U.S. Coast Guard photograph.

yachts, the methods of rescuing survivors from the water are many and varied, with the level of difficulty usually dependent on the height of the freeboard and the number of hands available to assist with recovery. The onus is therefore on the captain of such a craft to consider the problems and devise a standard rescue procedure for the vessel. Ideally, the captain should test the chosen technique in realistic conditions to prove its practicality. In some situations, an extraneous aid, such as a rescue net or similar means for parbuckling a cold or injured casualty on board, might be necessary.

Phase III: Post-Rescue

The most important cause of post-immersion death is hypoxia (shortage of oxygen), secondary to the aspiration of water—that is, the delayed effects of near drowning (see chapter 5). Many immersion victims who are rescued alive subsequently die. Many are even conscious and breathing spontaneously at the time of rescue, only to deteriorate later because of a progressive shortage of oxygen in the blood and tissues. It is imperative, therefore, that someone competent should check anyone who may have aspirated water to decide whether hospital screening is necessary. In general, those with any of the following signs and symptoms should be admitted to a hospital for X-ray and blood-gas analysis as soon as possible:

▸ Extraneous sounds (crackling and wheezing) in the chest on auscultation (or when an ear is applied to the chest wall)

▸ Breathing difficulty, especially if accompanied by cyanosis

Even those who are initially free from the preceding signs and symptoms should be periodically screened in the hours following an immersion incident in which they aspirated some water in case they develop so-called secondary drowning (chapter 5). Although it is a rare condition, it can catch the unwary, often with tragic results. When it manifests itself, immediate treatment in an intensive-care facility is necessary to prevent death.

The delayed effects of the low blood pressure suffered during rescue through the mechanisms described previously may also cause death in the immediate post-rescue period.

"Rewarming Collapse"

A less common cause of post-rescue death may be the relative shortage of circulating blood volume because of overzealous rewarming. This can occur when superficial extraneous rewarming dilates the blood vessels in the skin and superficial tissues, which had previously been cold vasoconstricted. This increases demand on both cardiac output and circulating blood volume. If uncompensated by a redistribution of blood from some other section of the vascular bed, blood pressure will fall. When body temperature is normal, the baroreceptors detect the onset of this fall in pressure and reflexly institute the appropriate compensatory adjustments to restore pressure to normal. In hypothermia, however, this regulating response is impaired or absent. Consequently, blood pressure will fall. The drop may be catastrophic, especially in those individuals with underlying coronary artery disease. Victims who are already fluid depleted at the time of rescue will be particularly susceptible to this problem if they are rewarmed too quickly.

Too Much Warmth Is a Bad Thing

"Everyone was conscious when taken out of the water, but many lost consciousness when taken into the warmth of the trawler. Nine (of 12) died soon after being rescued" (Capt. J. Downey, SS *Empire Howard*).

The fall in blood pressure, if marked, can reduce coronary blood flow to a level at which the heart muscle has insufficient blood supply to meet its requirements. This occurrence is especially likely in the middle aged and elderly, who may already have a reduced flow through disease-related narrowing of the coronary arteries, and in those who remain upright because this increases the possibility of postural hypotension. A fall in blood pressure may also produce harmful effects to the brain and kidneys.

Other Possible Causes for Post-Rescue Collapse

Another potentially fatal delayed effect in the post-rescue phase is hemorrhage, particularly one that is not obvious (for example, rupture of liver or spleen from trauma suffered during the initial fall into the water). The internal bleeding may not manifest itself until rewarming is nearing completion, when vascular beds are reopening and blood pressure is returning to near normal levels.

Infection of the lungs may be a late complication, particularly if an antibiotic-resistant organism causes it.

TREATMENT OF THE RESCUED

From the foregoing we can conclude that the principal medical problems likely to be encountered in immersion victims include near drowning, hypothermia, cardiovascular disorders, and trauma. Survivors rescued from survival craft may also be suffering from some or all of the foregoing, although they are likely to be more severely affected by hypothermia and less from near drowning. In addition, they may be suffering from cold injuries (chapter 10), dehydration and starvation (chapter 8), seasickness, and skin disorders (chapter 10). Any of the preceding may seriously compromise the survivors' ability to assist in their own rescue. Rescuers must operate on that assumption before locating the survivors and make suitable arrangements for assisted recovery.

Rescue

Rescuers should achieve recovery of the victim speedily without compromising the safety of the survivor or themselves. They should instruct survivors to do as little as possible to assist in their rescue. In some situations, air trapped in clothing, waterproof clothing in particular, may be critical in helping a victim without a life jacket to remain afloat. By moving about in an attempt to assist in the rescue, the victim may displace this air, lose buoyancy, and sink.

In general, rescuers should bring the victim inboard as quickly as possible. If wave splash threatens the victim's airway, rescuers should remove the person from the threat as quickly as possible by any means available. Alternatively, a survivor who has been in the water for some time and has kept the airway clear of the water (i.e., has a life jacket with splash guard) may be suffering from severe hypothermia. If circumstances permit, this casualty should be handled as if he or she were critically ill. With such victims, the speed of recovery is less important than careful removal from the water. In this situation, rescuers should try to recover the victim in as near a horizontal attitude as possible to assist circulation. This can be achieved using a rescue net or similar device to parbuckle the victim over the side or transom of the rescue boat (figure 11.6). Helicopter rescues are best carried out using a double strop or rescue basket (see figure 11.5 on page 262).

Regardless of which attitude they are recovered in, victims should be placed in a horizontal position or with legs slightly elevated as soon as possible after rescue.

Figure 11.6 Diagram of a casualty being parbuckled on board a semirigid inflatable boat.

Treatment of Immersion Survivors

As discussed previously, the most serious threat to life in immersion victims, and the condition in most urgent need of treatment on rescue, is near drowning. The primary aim of treatment of near-drowning victims is early correction of the oxygen deficit of the tissues and prevention of secondary complications. It is now generally accepted that early administration of oxygen to all near-drowning victims is essential, even to those who are fully conscious.

Immediate Management on Rescue

The major aim of immediate management at the rescue site is to ensure that both ventilation and circulation are adequate and provided by artificial means if required. In an inflatable or semirigid inflatable FRC, this will be more difficult to achieve than it would be on the firm structure of a ship, boat, yacht, or helicopter. Nevertheless, rescuers should immediately check the airway, clear it if necessary, and administer expired air ventilation (EAV) if the victim is not breathing. Thereafter, they should take steps to prevent any further heat loss from the body while in transit to the place where additional treatment will be undertaken.

Once the survivor is on board the rescue craft, rescuers should consider the optimum position in which to lay the victim to offset any potential problem with maintaining blood pressure. In an FRC, it is desirable to lay the casualty in a feet-forward, head-aft attitude; the elevation of the bow when under way will assist venous return. Conversely, in a helicopter the head should be forward and the feet aft.

Airway maintenance.

Some authorities still advocate draining the airway before commencing resuscitative maneuvers, whereas others recommend using abdominal thrusts *(Heimlich maneuver)*. The current consensus is that there is little point in delaying ventilation by trying to drain water from the lungs before commencing EAV. Experimentation has shown that neither gravitational drainage nor abdominal thrusts make any significant difference to pulmonary gas exchange or recovery rate when compared with standard cardiopulmonary resuscitation (CPR) techniques in near-drowning victims. Additionally, the maneuver itself may cause vomiting, delay initiation of more effective methods of ventilation, aggravate other injuries, cause confusion among rescuers (yet another protocol), and increase the risk of precipitating undesirable cardiac arrhythmia in the hypothermic casualty. The standard finger-sweep technique for clearing debris or vomit from the mouth is all that is necessary before commencing EAV. Any free water in the mouth or throat will probably already have drained during the rescue procedures.

Ventilation.

In the absence of spontaneous respiration on clearing the airway, the first-aid practitioner should give two breaths of mouth-to-mouth respiration. Such distension of the nerve stretch receptors in the lungs may cancel the effect of any vagal inhibitory action on the heart (sometimes a cause of sudden death) and restore satisfactory cardiac rhythm. First-aid practitioners should know that because of the increased stiffness of the lungs in near-drowning victims, the lungs will be more difficult to inflate. If the person performing mouth-to-mouth respiration has difficulty inflating the lungs, he or she should make sure that the airway is extended and clear. This practice will also prevent air from being forced into the stomach. If spontaneous respiration returns, the practitioner should place the victim in the recovery position and insulate to prevent further heat loss. Be watchful for regurgitation or vomiting during resuscitation or recovery; approximately 60 to 80 percent of near-drowning victims vomit during resuscitation. If available, oxygen (100 percent) should be administered, with bag and mask if necessary, to all near-drowning victims, regardless of their state of consciousness.

Airway clearance and the commencement of resuscitation will always take precedence over a speedy return to the shore or parent vessel. Spending a few minutes restoring a clear airway and reoxygenating the lungs through EAV immediately on rescue could prevent severe brain damage and may prove lifesaving. Rescuers should resist the desire to rush back to the mother vessel or shore until they complete a careful assessment of the casualty. Making such an assessment when the FRC is underway, especially at speed, is extremely difficult and in a rough sea is almost impossible. Therefore, rescue personnel should remain calm and carry out practiced routines slowly and deliberately rather than rush back to shore.

Once the FRC is underway it will be extremely difficult to make a good mouth-to-mouth or mouth-to-nose seal to carry out EAV effectively. A person is liable to damage teeth in trying to do so. Consequently, a flexible hose connecting the mouth of the resuscitator to an oronasal mask and incorporating a nonreturn valve makes the task easier. The device also frees a hand to steady oneself. It may be necessary for the operator of the FRC to slow or even stop at intervals to facilitate effective EAV. The delay incurred by an interrupted journey will prove less damaging to the victim than a three- or four-minute period of oxygen starvation of the brain during a rapid, nonstop return to the shore or parent vessel.

Cardiac arrest.

Conventional first-aid teaching recommends that external cardiac massage (ECM) be commenced in the absence of a pulse in the neck. Diagnosing cardiac arrest in an immersion victim in an FRC, however, can be extremely difficult. The inability to detect a pulse in the neck can result from factors such as boat movement, cold (numb) fingers in the rescuers, severe vasoconstriction, low blood pressure, and bradycardia (marked slowing of the heart rate) in the hypothermic victim. In addition, conducting effective ECM in an FRC that is under way is extremely difficult. For these reasons, unless someone witnessed the cardiac arrest, in the United Kingdom[1] it is considered inadvisable to commence ECM in these circumstances, even in the absence of a palpable pulse. Instead, it is better to wait until the victim has been returned to shore or the parent vessel, where a thorough assessment of the victim's condition can be made more easily and more effectively.

The rationale behind this unconventional advice is as follows. Commencing ECM in a severely hypothermic victim may precipitate ventricular fibrillation and cardiac arrest. On rescue, even if a pulse cannot be detected there is a possibility that the heart is still beating slowly but at sufficient speed and efficiency to meet the circulatory demands of the hypothermic body. If ECM is commenced and cardiac arrest results, it is then mandatory to perform effective ECM continuously to maintain viable circulation until electrical defibrillation can be performed. Because the chances of accurately diagnosing cardiac arrest in the first place and then carrying out effective ECM are remote, it is better to defer any such action until reaching the mother vessel. On the more stable platform, a better assessment can be made with the assistance of the necessary technical aids. Additionally, because in severe hypothermia the brain requires a greatly reduced amount of oxygen, a relatively short delay in commencing ECM puts the individual at less risk than commencing ECM and possibly inducing a cardiac arrest that may subsequently prove intractable to defibrillation. Finally, in the absence of the necessary advanced medical therapeutic and biochemical support, the effectiveness of ECM as a lifesaving measure diminishes rapidly with time from the onset of cardiac arrest. In most cases when ECM has proved successful in saving life, the cardiac arrest was witnessed, ECM was commenced almost immediately, and relatively rapid access to a defibrillator and critical-care medical facilities was available.

The remarkable case of resuscitation of a skier who fell through ice, described in chapter 5, may serve as a counterargument to this advice. The victim received full CPR both in the field and in the rescue helicopter for 1 hour and 50 minutes and eventually revived. Although it was a remarkable case and a great tribute to those providing her care, the role played by cardiac massage in her recovery is debatable because with a

---

1. Other countries may adopt different policies with regard to the initiation of ECM in such scenarios. The reader should consult national guidelines in this respect.

deep body temperature of 13.7 degrees Celsius (56.6 degrees Fahrenheit), her oxygen requirements would have been negligible.

Insulation.

Having established that the victim is breathing spontaneously, or commencing EAV if not, the next step is to insulate against further heat loss. This can be achieved by enclosing the survivor in a windproof and waterproof material such as polyethylene. It is important to cover the head, which is a major source of heat loss from the body in cold ambient environments. In such conditions, with wet hair and high levels of relative air movement, over 50 percent of body heat loss may occur through the head.

Treatment on the Parent Vessel

On returning to the parent vessel, a more careful examination should be carried out. Oxygen therapy should be commenced as soon as possible. If there is any evidence of near drowning (coughing, cyanosis, noisy chest, labored breathing, confusion and agitation), arrangements must be made for air evacuation to a hospital. In such cases, the provision of additional oxygen to the inspired air of someone breathing spontaneously, but with inadequate ventilation, will be insufficient to oxygenate the blood adequately. Additional pressure is usually required to force the air into the collapsed or fluid-filled terminal airways. This can be achieved only by administering gas under pressure. In ideal conditions, someone competent should induce relaxation (anesthesia) and perform intubation using appropriate drugs and equipment. In the absence of sophisticated equipment, EAV may supply lifesaving pressure that spontaneous breathing alone cannot achieve, but the patient must not fight against it.

If oxygenation of the blood remains inadequate, the level of consciousness will continue to deteriorate, and the victim will eventually die unless intubated. In the absence of a competent person to intubate, first-aid practitioners should consider alternatives. The use of a positive end expiratory pressure (PEEP) valve, a laryngeal mask airway (LMA), or a "Combitube" are possible contenders as last-ditch emergency measures for unconscious patients. These are specialized items of equipment, however, and training in their use is necessary. The patient must be unconscious for either the LMA or "Combitube" to be inserted.

In an unconscious near-drowning victim, a decline in pulse rate and falling blood pressure usually indicate the onset of terminal circulatory failure.

Rewarming

After a thorough examination, immersion victims who are suffering solely from uncomplicated hypothermia should be assessed with regard to their suitability for one of the following rewarming regimes:

▶ Rapid external (active rewarming)

▶ Slow spontaneous (passive rewarming)

▶ Assisted passive rewarming

Active rewarming.

In general, immersion victims who are moderately cold (rectal temperature greater than 34 degrees Celsius; 93 degrees Fahrenheit), shivering, and fully conscious after a relatively short immersion may be rapidly rewarmed without risk. A bath of hot water (40 degrees Celsius; 104 degrees Fahrenheit) under supervision is the best way to achieve this. Hot showers are less effective as a rewarming medium and have the added risk of inducing *rewarming collapse*. If showers are the only means available for rapid rewarming, someone must constantly supervise victims and warn them that if they feel even slightly dizzy, they should say so. They should then be removed from the shower to continue rewarming by the slow passive method.

Active rewarming has the advantage that it rewarms the victim quickly and thus rapidly restores the feeling of well-being, thereby reducing the stress suffered by the victim. An added advantage is that it inhibits or reduces the intensity of shivering and thereby reduces the workload of the heart, besides having other biochemical and circulatory benefits. Some recommend leaving the limbs out of the hot water during the rewarming process to reduce the potential for cold blood returning to the central circulation and intensifying the magnitude of the afterdrop. Clearly, our current understanding of the mechanism of the afterdrop obviates this requirement.

The major danger associated with rapidly rewarming mildly hypothermic survivors is that they may stay too long in the comfort of the bath and suffer rewarming collapse on leaving the water. They should leave the bath when they feel they have rewarmed sufficiently and certainly before they start to sweat. If someone is monitoring the ECG or pulse rate, an increase in heart rate will indicate that rewarming should end. On leaving the water and putting on dry clothing, survivors should lie down on a bunk for a least an hour to continue their recovery. During this period the survivors may consume hot, sweet drinks.

In a major disaster many casualties may need rewarming at the same time. In such situations passive rewarming may be the only alternative to hot showers. If it is deemed that showers are appropriate, then in the absence of individual supervision, victims should be advised to sit on the deck beneath the shower to reduce the risk of rewarming collapse, which could result in fainting and additional traumatic injury. Again, care should be exercised to ensure that victims do not overheat. After showering, victims should lie down for at least an hour to complete their recovery. Supervision should continue during this time. Before leaving the sick bay, victims should be reexamined to check, in particular, for any evidence of water aspiration.

Passive rewarming.

Hypothermic survivors from life craft or immersion victims who are semi-conscious or unconscious are best rewarmed passively. Those providing aid should gently remove the victim's wet garments with minimal disturbance. If necessary, the clothes should be cut off, particularly in unconscious victims. The victim should then be placed in a sleeping bag or wrapped in blankets to prevent further heat loss. The head should be insulated. The body's own metabolism initiates rewarming through shivering, which generates heat internally, thereby rewarming the body from the inside out.

Passive rewarming is the method of choice in the out-of-hospital management of all severely hypothermic cases (body temperature less than 34 degrees Celsius; 93 degrees Fahrenheit), especially those who are semi-conscious or unconscious. The slow rate of rewarming (0.5 to 1 degree Celsius [0.9 to 1.8 degrees Fahrenheit] per hour) may prolong the victim's suffering (shivering), compared with rapid rewarming, and it also prolongs the time that the heart is required to operate at an increased workload, both to support the continued high blood flow to shivering muscle and to work against vasoconstricted peripheral blood vessels. Nevertheless, passive rewarming is the safest method of rewarming profoundly hypothermic patients. Slow rewarming has the advantage of enabling physiological regulatory processes to recover as the temperature of the brain slowly increases. Thus, provided the rate of rewarming is controlled at the recommended level, this technique reduces the threat of rewarming collapse and other complications. Again, the victim may drink hot sweetened drinks during recovery after regaining consciousness.

Assisted passive rewarming.

Although for very cold individuals passive rewarming is often the safest and easiest technique, it relies on the body's ability to generate heat through shivering. Therefore, this method is ineffective if severe injury, profound hypothermia (less than 30 degrees Celsius; 86 degrees Fahrenheit), or intoxication inhibit shivering.

When shivering is absent, and body temperature is near a lethal level (less than 30 degrees Celsius; 86 degrees Fahrenheit), or it continues to fall 30 minutes after passive rewarming methods have been instituted, extraneous heating may be necessary. In such cases a moderately warm hot-water bottle (or other method of controlled local application of heat, such as an electrically heated blanket with good thermal control) applied to the abdomen may help reverse the fall in body temperature. Care must be taken to ensure that the temperature of the heat source is not too hot. With impaired circulation to the skin, burning can occur at temperatures that a normothermic individual perceives as comfortably warm. If in doubt, the bottle should be wrapped in a towel or similar material to prevent it from coming into direct contact with the skin of the cold abdomen. As long as the temperature of the bottle is several degrees above that of the rectal temperature, heat will flow into the body and contribute to the rewarming process. Moderately warm bottles will require frequent replacement.

Two popular methods of assisted rewarming are *inhalation rewarming* and *forced-air rewarming.* Theoretical and laboratory-based evaluations have demonstrated that inhaling heated, humidified air contributes little heat to the body. Thus, we can probably more accurately describe this method as providing respiratory insulation, in that it prevents further respiratory heat loss in a cold environment. Having said this, anecdotal accounts from those who have used inhalation rewarming in the field describe remarkable improvements in the condition of profoundly cold victims. It is possible that although breathing hot, moist air does not contribute much heat to the whole body, it may selectively warm the back of the throat and base of the brain, an area important for the maintenance of respiratory and cardiac function.

Forced-air rewarming involves blowing hot air through a containing cover over the surface of the body. The danger of this increasingly popular technique is that it may rewarm individuals too quickly and provoke rewarming collapse. But if a victim is unable to generate sufficient metabolic heat (evidenced by absence of shivering, slow heart rate and respiration, and continued cooling), such a technique may be considered. The difficulty of measuring deep body temperature, heart rate, and respiration in adverse field conditions, coupled with the potential complication (hypotension) of using assisted rewarming, make the decision to use this technique difficult.

Unconscious patients must be placed in the "unconscious (recovery) position" throughout the rewarming procedure to ensure that the victim maintains an unobstructed airway. Frequent monitoring of deep body temperature, although difficult to accomplish, is necessary to ensure that rewarming is proceeding at a satisfactory rate (0.5 to 1 degree Celsius [0.9 to 1.8 degrees Fahrenheit] per hour). Faster rewarming can precipitate rewarming collapse (as indicated by an increase in pulse rate), so the rate of rewarming should be controlled by judicious adjustment of the insulation around the body.

On regaining consciousness, the patient should receive reassurance and may consume hot, sweet drinks. Ideally, patients should be in the company of fellow survivors and discuss their experiences if they so desire. When rewarming is complete, a medical authority should provide advice about further care of the casualty, including the possible need for counseling.

Other Coexisting Conditions

Coexisting conditions other than near drowning, hypothermia, dehydration, and starvation may be present in survivors. These include seasickness, cold injuries, heat-related illnesses, contamination with oil, and osmotic diarrhea. Chapter 10 discusses these conditions.

Life-Craft Survivors

Inadequately protected castaways rescued from survival craft in temperate or subarctic environments may be suffering from hypothermia, even after voyages of relatively short duration. In the colder latitudes, they may also be suffering from cold injuries (chapter 10). Even in temperate climatic regions, castaways will probably succumb to the effects of cold long before they yield to the effects of dehydration or malnutrition. Life-craft occupants who manage to survive for some time in these cold conditions will probably be suffering from the chronic effects of hypothermia. During rewarming, dehydration, which may not manifest itself until normal circulation is restored, may complicate their recovery. This form of dehydration is not necessarily due to inadequate fluid intake but is more likely the consequence of the physiological adjustments to circulating blood volume. These adjustments result from the intense cold vasoconstriction and the compensatory excretion of fluid, or *cold diuresis.* Additionally, kidneys at low temperature have limited ability to reabsorb fluid; this increases urine volume and, with time, further depletes body fluid stores.

Such fluid losses may not cause any outward manifestations of dehydration in a hypothermic survivor at the time of rescue. On rewarming, however, as circulation recovers, fluid deficit may cause a substantial, or even fatal, fall in blood pressure *(hypotension).*

Rewarming of such survivors should follow the passive or assisted passive protocols detailed earlier to avoid complications. But starving survivors may have insufficient blood glucose to facilitate shivering (chapter 8). If the survivor is conscious, or even semiconscious, some sugar may be given to kick start shivering into action. Should the rate of rewarming exceed the optimum (0.5 to 1 degree Celsius; 0.9 to 1.8 degrees Fahrenheit per hour), the level of insulation can be modified to control it.

In long-duration survival voyages in warmer climates, survivors are likely to have experienced both dehydration and malnutrition, although some may be suffering from moderate levels of hypothermia, cold injuries, or skin ulceration (chapter 10).

On rescue, those providing aid must exercise caution in recovering survivors. Even though survivors may not appear to be dehydrated or malnourished, they will need assistance in recovery into the rescue vessel. After a period of relative inactivity in a semirecumbent attitude in a life raft, they are likely to be physically weak, experience difficulty with locomotion, and be particularly susceptible to postural hypotension. If possible, rescuers should recover survivors into a rescue boat first, and then retrieve the boat by davit. This procedure will permit the maintenance of a horizontal attitude throughout. After rescue, such victims should remain in a horizontal attitude while someone assesses their general condition. At this stage the victims may drink some sips of water, but they should not swallow large quantities at once. Caution should be exercised in attempting to reverse either dehydration or malnutrition rapidly. A light diet with minimal roughage containing sugar, essential proteins, and small amounts of fat is recommended for a few days until the stomach and gut have recovered and returned to normal physiological activity. Initially, a diet composed predominantly of beaten eggs, sugar, and milk (or water) is sufficient. Survivors should consume it slowly, frequently, and in small quantities. After a day or two the survivors may resume a normal diet, with added vitamins.

Rescued survivors may suffer residual psychological trauma, experience flashbacks, and encounter problems sleeping.

Post-Traumatic Stress

Many shipwreck survivors are later haunted by the memory of the sights and sounds of loved ones or close acquaintances in death throes. Most will have an overwhelming sense of guilt that they survived while others died. Thoughts that they could have, and perhaps should have, done more to help others will reinforce these feelings. This irrational sense of shame will recur, and survivors will overlook the impossibility of helping others when they were themselves overcome. They may be incapable of analyzing the situation dispassionately and understanding that they used all their physical capacity to save themselves, leaving none to help others. Finally, if relatives or close friends died in the incident, the survivor may lack traditional sources of support and consolation.

The stressful period of post-trauma reflection (*post-traumatic stress*) begins when the victims have been rescued or even earlier for those in the relative safety of their survival craft. Although physically safe, they now must come to terms with the aftermath of the disaster. Feelings of guilt that they have survived when others have not create mental anguish, frequently resulting in dreams incorporating flashbacks. Sleeplessness, aggression, other behavioral changes, and dependency problems often develop. These behavioral changes strain relationships and frequently end in loss of employment, divorce, or even suicide. The term given to this chronic condition is *post-traumatic stress disorder* (PTSD). It is important to be aware that such survivors may later suffer from PTSD.

Experts generally agree that the best treatment for post-traumatic stress and its possible progression to PTSD is group counseling with fellow survivors, not outsiders (although the leader of the discussion group will probably have to be someone suitably trained). The leader should encourage all to talk freely but should discourage recrimination. Prevention of insomnia in the immediate post-rescue period is important. Alcohol or, if available, an appropriate pharmaceutical preparation prescribed by a doctor may help prevent sleeplessness. To avoid later flashbacks, survivors should be encouraged not to bottle up any feelings of guilt.

In conclusion, those who are preparing for and protecting themselves from the physiological and medical threats associated with survival at sea should also be aware of some of the associated psychological threats.

Chapter Summary and Recommendations

> Wearing a securely anchored safety harness at all times when on the upper deck in bad weather will significantly reduce the chances of becoming a man-overboard victim.

> Approximately 20 percent of rescued survivors die during or shortly after rescue (circumrescue collapse).

> Anecdotal evidence suggests that three phases during the rescue process are associated with particular risks:

 a. Pre-rescue—just before rescue

 b. During rescue—during or immediately after removal from the water

 c. Post-rescue—following rescue

> Most deaths result from drowning.

> The remainder of the deaths result not from the afterdrop but from the following:

 a. A collapse in blood pressure associated with

  1. prolonged exposure,

  2. hypothermia,

  3. mode of rescue (posture, activity),

  4. loss of the hydrostatic assistance to venous return,

  5. reintroduction of the full effects of gravity,

  6. decreased circulating blood volume (hypovolemia),

  7. increased blood viscosity, or

  8. dulled baroreceptor reflex responses.

b. An increase in the work rate of a cold heart when aiding in one's own rescue (e.g., climbing up ladders in high-sided ships), especially in people with preexisting cardiac problems

c. Excessively rapid rewarming (rewarming collapse)

d. Hemorrhage from an internal injury aggravated during the rescue process, or later as blood pressure recovers

▶ It follows from the preceding that the transition from water to air during rescue from water at any temperature is likely to be less traumatic if subjects are lifted horizontally;

a. survivors whose airways are not under threat of aspiration should be rescued with care, preferably horizontally, and handled as if they were critically ill, however

b. survivors still in the water and whose airways are under threat of aspiration of water should be rescued as quickly as possible by whatever means are available.

▶ Once on board the rescue craft, the victim should be placed in the optimum position to offset any potential problem in maintaining blood pressure. In an FRC, it is desirable to lay the casualty in a feet-forward, head-aft attitude; in a helicopter the head should be forward and the feet aft. A competent person can then assess the victim's general condition and begin appropriate first-aid action.

▶ The major aim of immediate management at the rescue site is to ensure that the airway is clear and ventilation (EAV) is provided if required.

▶ Because the most important cause of post-immersion death is hypoxia (shortage of oxygen) secondary to the aspiration of water, near-drowning victims should receive oxygen as soon as a clear airway has been established.

▶ All near-drowning survivors should receive medical attention as soon as possible.

▶ Care should be exercised to ensure that those who require resuscitation do not aspirate vomit.

▶ Cold survivors must be protected from further heat loss, in particular evaporative heat loss and heat loss through forced convection.

▶ Rewarming regimes can be broadly categorized as

a. rapid external (active rewarming),

b. slow spontaneous (passive rewarming), and

c. assisted passive rewarming.

▶ The decision about which technique to use is difficult and depends on circumstances. In general, passive rewarming at a rate of 0.5 to 1 degree Celsius (0.9 to 1.8 degrees Fahrenheit) per hour is the safest option.

▶ Standard operating procedures for rescue, recovery, and treatment should exist, and those with responsibility in this area (e.g., crews of ships with FRC) should practice regularly.

▶ All boat skippers should consider the problem of recovering someone from the water and know how to achieve it.

▶ Survivors from survival craft may also have to be recovered in a horizontal attitude.

▶ Survivors may suffer from post-traumatic stress at some later stage. The best treatment for post-traumatic stress and its possible progression to PTSD is group counseling with fellow survivors.

Appendix: Conversion Factors

Table A1   Volumetric Conversions

To convert from U.S./imperial to metric	Multiply by	To convert from metric to U.S./imperial	Multiply by
Fluid ounces (imperial)	28.413063	Millilitres (ml)	0.035195
Fluid ounces (U.S.)	29.5735	Millilitres (ml)	0.033814
Pints (imperial)	0.568261	Liters (L)	1.759754
Pints (U.S.)	0.473176	Liters (L)	2.113377
Quarts (imperial)	1.136523	Liters (L)	0.879877
Quarts (U.S.)	0.946353	Liters (L)	1.056688
Gallons (imperial)	4.54609	Liters (L)	0.219969
Gallons (U.S.)	3.785412	Liters (L)	0.364172

Table A2   Weight Conversions

To convert from U.S./imperial to metric	Multiply by	To convert from metric to U.S./imperial	Multiply by
Ounces	28.349523	Grams (g)	0.035274
Pounds	0.453592	Kilograms (kg)	2.20462

Glossary

acclimation—Alterations in the response of the body after repeated exposure to a stimulus (e.g., heat or cold).

acute—An adjective used to describe a condition of short onset and duration.

adrenaline (epinephrine)—A catecholamine (one of the hormones involved in the response to stress—flight-or-fight response).

aerobic exercise—Exercise in which oxygen is used in the production of energy; as opposed to anaerobic exercise, in which no oxygen is used or available (*see also* maximum aerobic capacity).

afterdrop—The continued fall in deep body temperature seen during the early stage of rewarming in hot water after body cooling. Afterdrop should not be confused with the temperature fall associated with continued cooling after rescue because of the provision of inadequate insulation or insufficient extraneous heat.

alveolar membrane—The thin cell membrane separating air from blood in the alveoli.

alveoli (alveolus)—The terminal air sacs in the lungs where the gas interface with blood takes place.

amnesia—Loss of memory.

anaerobic—Exercise in which no oxygen is used or available for the production of energy.

anhydrated—Totally without water, in contrast to being short of water when dehydrated.

anoxia—Without oxygen, in contrast to a shortage of oxygen when hypoxic.

apnea—A temporary cessation of breathing.

asystole—When the heart no longer beats and the ECG (EKG) is flat.

autoregulation—A process in which feedback mechanisms in the body sense alterations from the normal resting state and reflexly trigger the appropriate physiological adjustment to return to the status quo.

baroreceptors—Nerve sensors in the large blood vessels near the heart that are sensitive to pressure changes in the blood. They respond reflexly to rises or falls in blood pressure and make the necessary autoregulatory adjustments to the circulation to restore normality.

bellows effect—The intermittent exchange of environmental air with the air from the microenvironment beneath clothing, usually through limb movement (during exercise) when clothing apertures are open.

boundary layer—Term used to describe the layer of air or water molecules immediately adjacent to the external surface of the skin or any physical surface of a body in air or water.

brain stem—The portion of the brain connecting it to the spinal cord and containing the important cardiac and respiratory centers.

bradycardia—Slowing of the heart rate to less than 50 beats per minute.

capillaries—The microscopic blood vessels connecting minute arteries to veins.

carbon dioxide—A gas produced in the body as a product of metabolism.

cardiac output—The volume of blood pumped out by the heart (usually expressed in milliliters per minute or liters per minute).

cardiovascular—Pertaining to the heart and circulatory systems of the body.

catabolism—The breakdown by the body of its own tissues for energy.

catecholamines—Chemicals (hormones, e.g., adrenaline and noradrenaline) released into the blood stream in response to stress.

cell membrane—The wall or boundary of a cell through which nutrients and fluids pass to and from the cell.

cerebral—Pertaining to the brain.

cerebral ischemia—Insufficient blood flow to the brain.

chronic—An adjective used to describe a condition of long duration.

circumrescue collapse—Collapse just before, during, or shortly after rescue. Includes post-rescue collapse and rewarming collapse.

CIVD—Cold-induced vasodilatation. Vasodilatation of cooled blood vessels resulting from cold-induced paralysis of the muscle of the blood-vessel wall.

clo—The arbitrarily agreed-upon unit used to describe the insulation of garments. One clo is equivalent to the amount of insulation required to keep a seated subject comfortable in air at a temperature of 21 degrees Celsius (70 degrees Fahrenheit), less than 50 percent relative humidity, and less than 0.1 meters per second (0.22 miles per hour) of air movement (about the insulation provided by a business suit). See also footnote 1, chapter 3.

CNS—Central nervous system (the brain and spinal cord).

cognitive—Involving conscious mental processes.

cold constrictor fibers—Nerves that produce vasoconstriction when stimulated.

cold diuresis—An increase in urine production by the body during cold exposure.

cold injury—Tissue injuries resulting from short-duration exposure to severe cold or prolonged exposure to less severe cold ambient conditions.

cold receptors—Nerve endings in the body, particularly the skin, that sense and respond to a fall in their temperature.

cold shock—The response of the body to sudden cooling of the skin, usually on immersion in cold water, characterized by rapid breathing and heart rate and an increase in blood pressure.

CPR—Cardiopulmonary resuscitation.

cutaneous—Pertaining to skin.

cyanosis—A blue discoloration of the lips and nail beds, possibly indicating a shortage of oxygen.

dehydration—A depletion of body water.

diaphragm—The dome-shaped respiratory muscle separating the lungs from the abdominal contents.

diuresis—An increase in urine production.

diving reflex or diving response—A reflex found in diving animals that enables them to hold their breath while underwater for a longer period than in air.

EAV—Expired air ventilation; the respiratory component of CPR.

ECM—External cardiac massage; the cardiac component of CPR.

edema—A local accumulation of fluid in tissues.

energy balance—The balance between food consumed and energy expended by the body.

epinephrine—*See* adrenaline.

EPIRB—Emergency position indication rescue beacon.

extracellular fluid (ECF)—The body fluid not within the boundary of the cell *(cell membrane)*.

FCI—Freezing cold injury (e.g., frostbite).

FRC—Fast rescue craft

frostbite—Cold injury resulting from the freezing of tissue fluid (*see* FCI).

frostnip—Freezing of the superficial skin, classically of the cheeks, tip of nose, ears, or fingertips.

gluconeogenesis—The biochemical process by which glucose is synthesized from fat and protein in the body (mainly in the liver and kidney).

glycogen—The form in which sugar (glucose) is stored in the body.

heat exhaustion—Early stage of heat illness characterized by dizziness, faintness, tiredness, headache, and possibly nausea and vomiting.

heat stroke—Severe, often terminal, form of heat illness usually characterized by hot, dry skin.

homeostasis—The physiological process by which the internal systems of the body (e.g., blood pressure, temperature, acid-base balance, etc.) are maintained at equilibrium despite variations in external conditions.

homeotherms—A genus of animals, including humans, that try to control their deep body temperature within a given range regardless of the ambient temperature.

hormones—Chemicals released into the blood stream that produce a response in other organs in the body.

hydrocution—The name sometimes used to describe sudden, unexplained death after immersion in previously fit young people.

hydrophobic—Materials that are resistant to wetting.

hydrostatic pressure—The pressure exerted on a body underwater by the density of the surrounding water.

hyoscine—Drug used in the prevention and treatment of seasickness.

hypernatremia—Excess sodium (salt) in the body.

hypertension—High blood pressure.

hyperthermia—A raised deep body temperature.

hyperventilation—Rapid breathing, at a rate above that necessary for metabolic requirements.

hypoglycemia—Reduced blood-sugar level.

hypotension—Low blood pressure.

hypothalamus—An area in the brain that is particularly concerned with body temperature regulation.

hypothermia—A deep body temperature more than 2 degrees Celsius (3 degrees Fahrenheit) below normal (i.e., less than 35 degrees Celsius; 95 degrees Fahrenheit).

hypovolemia—Reduced circulating blood volume.

hypoxia—Shortage of oxygen.

immersion foot—Non-freezing cold injury of the feet from protracted immersion in cold water.

inflammation (inflammatory response)—The localized response of the body to injury, resulting in pain, swelling, heat, redness, and impairment of function.

intracellular fluid—The fluid contained within the cells of the body.

intubate—The insertion of a tube in the airway to assist with breathing.

laryngospasm—Muscle spasm of the throat that can obstruct breathing.

macroclimate—The general environmental climate outside the clothing or shelter.

maximum aerobic capacity ($\dot{V}O_2max$)—The limit of aerobic exercise. Exercise beyond this level can be achieved only by anaerobic means and is thus limited in duration. Fit people have a high aerobic capacity

metabolic rate—A measure of the level of metabolism taking place.

metabolism—The chemical activity within the body necessary for life.

microclimate—The climatic zone adjacent to the skin inside the clothing.

microenvironment—*See* microclimate.

narcotic—Sleep inducing.

near drowning—Survival, at least temporarily, after aspiration of fluid into the lungs.

NFCI—Non-freezing cold injury (e.g., immersion foot).

norepinephrine (noradrenaline)—A catecholamine (one of the hormones involved in the response to stress—flight-or-fight response).

normothermic—Normal body temperature.

osmotic pressure—The absorptive pressure exerted by a solution containing more solute molecules on one with a lesser concentration of solute molecules, across a semipermeable membrane (e.g., a cell wall).

paresthesia—Diminished sensation usually accompanied by "pins and needles."

pathophysiology—The abnormal physiology associated with tissue damage or injury.

PEEP—Positive end expiratory pressure. The term used to describe artificial maintenance of a positive pressure in the lungs to prevent their collapse at the end of expiration.

peripheral blood flow—The blood flow to the outer regions of the body.

PFD—Personal flotation device.

physiology—The science of the function of living organisms and their component parts.

placebo—A substance that has no medicinal effect, for example, a sugar tablet used in drug trials to control for the psychological effect when testing the true effectiveness of a pharmacological product.

post-rescue collapse—Collapse, often leading to death, during or shortly after rescue.

postural hypotension—A fall in blood pressure associated with adopting an upright posture.

potable—Safe to drink.

pulmonary—Pertaining to the lungs.

respiratory tract—The air passages to the lungs.

reverse osmosis—The reversal of the normal flow of fluid down an osmotic gradient by use of an artificially applied greater pressure.

rewarming collapse—Collapse during rewarming after protracted whole-body cooling.

subcutaneous fat—The layer of fat beneath the skin.

surfactant—An oily, fluidlike substance produced in the small airways and alveoli to prevent lung collapse through surface tension.

systemic blood pressure—The pressure within the arteries in the general circulation.

TEMPSC—Totally enclosed motor-propelled survival craft.

thermoneutral—An environmental temperature in which deep body temperature remains stable without evoking the stressful thermoregulatory response of sweating or shivering.

thermoreceptors—Nerve endings that are sensitive to either heat or cold found throughout the body.

thermoregulation—The regulation of body temperature, which may be achieved by either physiological or behavioral means, or more usually through a combination of both.

thermoregulatory zone—The environmental temperature range within which the naked body can regulate deep body temperature by adjusting blood flow alone.

TPAs—Thermal protective aids. Supplementary items of clothing or insulating devices to help conserve body heat of survivors in survival craft.

vagal inhibition—Slowing (possible stoppage) of the heart resulting from stimulating the vagus nerve.

vagus nerve—The 10th cranial nerve.

vapor permeable—Permits the passage of moisture vapor.

vascular bed—The blood vessels in a region of the body, for example, muscular vascular bed.

vasoconstriction—The active reduction in the diameter of blood vessels that results in the reduction, or even cessation, of blood flow to the tissues supplied by them.

vasodilatation—The increase in the diameter of blood vessels that facilitates an increase in blood flow locally.

venous pressure—The pressure exerted by the blood within the veins.

V.F.—Ventricular fibrillation. A state in which the heart muscle fibers contract in a chaotic, uncoordinated manner, preventing the heart from pumping effectively.

VHF—Very high frequency radio waves.

V̇O₂max—*See* maximum aerobic capacity.

warm receptors—Nerve endings in the body that sense and respond to a rise in their temperature.

wave splash—The term used to describe the intermittent breaking of waves over the head and face of an immersed person. It also encompasses the intermittent welling of water around the neck of a person wearing a life jacket (PFD) if buoyancy is insufficient to lift the airway clear of the surface at all times.

Bibliography

Allan, J.R. 1983. Survival after helicopter ditching: A technical guide for policy makers. *International Journal of Aviation Safety* 1:291–296.

Alexander, L. 1945. *The treatment of shock from prolonged exposure to cold, especially in water.* Combined Intelligence Objectives Sub-committee A.P.O. 413 C105, Item No. 24. HMSO. London.

Anderson, G.S., Meneilly, G.S., and I.B. Mekjavic. 1996. Passive temperature lability in the elderly. *European Journal of Applied Physiology* 73:278–286.

Bailey, M., and M. Bailey. 1974. *117-Days Adrift.* London: Cox & Wyman.

Barnett, P.W. 1962. *Field tests of two anti exposure assemblies.* Arctic Aerospace Laboratories Report No. AAL-TDR-61-56.

Bender, D.A. 1993. *Introduction to nutrition and metabolism.* King's Lynn and Guildford, England: Biddles Ltd.

Bennet, G. 1983. *Beyond endurance. Survival at the extremes.* London: Secker & Warburg.

Bennett, G.H., and R. Bennett. 1999. *Survivors: British merchant seamen in the Second World War.* London and Rio Grande, Ohio: Hambledon Press.

Bolte, R.G., Black, P.G., Bowers, R.S., Kent-thorn, J., and H.M. Correli. 1988. The use of extracorporeal rewarming in a child submerged for 66 minutes. *Journal of the American Medical Association* 260:377–379.

Bombard A. 1953. *The Bombard story.* London: Andrew Deutsch.

Bradish, R.F., Everhart, M.W., McCord, W.M., and W.J. Witt. 1942. Some physiological aspects of the use of seawater to relieve dehydration. *Journal of the American Medical Association* 9:683–685.

British Medical Association. 1964. Accidental hypothermia in the elderly. *British Medical Journal* 2:1255–1258.

Brooks, C.J. 1986. Ship/rig personnel abandonment and helicopter crew/passenger immersion suits: The requirements in the North Atlantic. *Aviation Space & Environmental Medicine* 57:276–282.

Brooks, C.J. 1995. *Designed for life: Lifejackets through the ages.* Mustang Engineered Technical Apparel Corp. Canada: Hemlock Printers Ltd.

Bullimore, T. 1998. *Saved.* London: Warner (Little, Brown & Company).

Burton, A.C. 1935. The average temperature of the tissues of the body. *Journal of Nutrition* 9:261–264.

Burton, A.C., and O.G. Edholm. 1955. *Man in a cold environment.* London: Edward Arnold.

Callahan, S. 1987. *Adrift—Seventy-six days lost at sea.* New York: Ballintine Books.

Carlson, L.D., and A.C.L. Hsieh. 1974. Temperature and humidity. In *Environmental physiology,* ed. N. Balfour Slonim. St. Louis: Mosby.

Cooper, K.E., Martin, S., and P. Riben. 1976. Respiratory and other responses in subjects immersed in cold water. *Journal of Applied Physiology* 40:903–10.

Conn, A.W., Miyassaka, K., Katayama, M., Fujita, M., Orima H., Barker, G., and D. Bohn. 1995. A canine study of cold water drowning in fresh versus salt water. *Critical Care Medicine* 23:2029–2036.

Critchley, M. 1943. *Shipwreck survivors: A medical study.* London: J&A Churchill Ltd.

Currie, James. 1798. Appendix II on the treatment of shipwrecked mariners in *The effects of water cold and warm as a remedy in fever.* London: Cadell & Davies.

Davis, D. 1992. *Aberdeen Press & Journal.* 16 March.

Eberwein, J. 1985. The last gasp. *U.S. Naval Institute Proceedings.* July:128132.

Edwards, B. 1990. *The grey widow maker.* London: Robert Hale Ltd.

Epstein, M. 1978. Renal effects of head-out immersion in man: Implications for understanding of volume homeostasis. *Physiology Reviews* 58:529–581.

Estonia. 1997. *The final report on the capsizing on 28 September 1994 in the Baltic Sea of the ro-ro passenger vessel MV Estonia.* The Joint Accident Investigation Commission of Estonia, Finland and Sweden. Helsinki: Edita Ltd.

Faith, N. 1998. *Mayday.* London: Macmillan.

Finkel, M. 1999. When hell freezes over. *Readers Digest.* April:122–130.

Foy, H., Altman, A., and M. Kondi. 1942. Thirst at sea—Seawater enemas. *South African Medical* Journal 6:113–115.

Franklin, J. 1868. *The life of Sir John Richardson.* London: Longmans, Green.

Fuller, R.H. 1963. Drowning and the post immersion syndrome: A clinicopathological study. *Military Medicine* 128:22–36.

Gale, E.A.M., Bennett, T., Green, H.J., and I.A. MacDonald. 1981. Hypoglycaemia, hypothermia and shivering in man. *Clinical Science* 61:463–469.

Gagge, A.P., Burton, A.C., and H.C. Bazatt. 1941. A practical system of units for the description of heat exchange of man with his thermal environment. *Science NY* 94:428–430.

Gilbert, M., Busund, R., Skagseth, A., Nilsen, P.A., and J.P. Solbo. 2000. Resuscitation from accidental hypothermia of 13.7°C with circulatory arrest. *Lancet* 355:375–376.

Gilley, W.O.S. 1850. *Narratives of shipwrecks of the Royal Navy between 1793–1849.* London: John W. Parker.

Glaser, E.M., and G.R. Hervey. 1951. Swimming in very cold water. *Journal of Physiology* 115:14P.

Glaser, E.M., and R.A. McCance. 1959. The effect of drugs on motion sickness produced by short exposures to artificial waves. *Lancet* 1:853–863.

Golden, F.St.C. 1973. Death after rescue from immersion in cold water. *Journal of the Royal Naval Medical Services* 59:5–8.

Golden, F.St.C. 1973. Recognition and treatment of immersion hypothermia. *Proceedings of the Royal Society of Medicine* 66:1058–1061.

Golden, F.St.C. 1976. Hypothermia: A problem for North Sea industries. *Journal of the Society of Occupational Medicine* 26:85–88.

Golden, F.St.C. 1977. *RN 25 man liferaft cold weather habitability trials 1975–1977.* Institute of Naval Medicine Report No. 14/77.

Golden F.St.C. 1983. Rewarming. In *The nature and treatment of hypothermia,* ed. R.S. Pozos and L.E. Wittmers, 194–208. Minneapolis: University of Minnesota Press.

Golden, F.St.C., Hampton, I.F.G., Hervey, G.R., and A.V. Knibbs. 1979. Shivering intensity in humans during immersion in cold water. *Journal of Physiology* 290:48P.

Golden, F.St.C., and P.T. Hardcastle. 1982. Swimming failure in cold water. *Journal of Physiology* 330:60–61.

Golden, F.St.C., and G.R. Hervey. 1977. The mechanism of the afterdrop following immersion hypothermia in pigs. *Journal of Physiology* 272:26–27.

Golden, F.St.C., and G.R. Hervey. 1981. The "afterdrop" and death after rescue from immersion in cold water. In *Hypothermia ashore and afloat,* ed. J.A. Adam. Aberdeen University Press, Aberdeen.

Golden, F.St.C., Hervey, G.R., and M.J. Tipton. 1991. Circum-rescue collapse: Collapse, sometimes fatal, associated with the rescue of immersion victims. *Journal of the Royal Naval Medical Service* 77:139–149.

Golden, F.St.C., and M.J. Tipton. 1987. Human thermal responses during leg-only exercise in cold water. *Journal of Physiology* 391:399–405.

Golden, F.St.C., and M.J. Tipton. 1988. Human adaptation to repeated cold immersions. *Journal of Physiology* 396:349–363.

Golden, F.St.C., Tipton, M.J., and R.C. Scott. 1997. Immersion, near-drowning and drowning. *British Journal of Anaesthesia* 79:214–225.

Goode, R.C., Duffin, J., Miller, R., Romet, T.T., Chant, W., and A. Ackles. 1975. Sudden cold water immersion. *Respiration Physiology* 23:301–310.

Gooden, B.A. 1992. Why some people do not drown; Hypothermia versus the diving response. *The Medical Journal of Australia* 157:629–632.

Goss, Pete. 1998. *Close to the wind.* London: Headline Book.

Grosse-Brockoff, F. 1946. Pathology and therapy in hypothermia. In *German aviation medicine World War II,* 828–842. Surgeon General USAF, Department of the Air Force.

Haight, J.S.J., and W.R. Keatinge. 1973. Failure of thermoregulation in the cold during hypoglycaemia induced by exercise and alcohol. *Journal of Physiology* 229:87–97.

Hall J.F., and J.W. Polte. 1956. Effect of water content and compression on clothing insulation. *Journal of Applied Physiology* 8:539–545.

Hardwick, R.G. 1962. Two cases of accidental hypothermia. *British Medical Journal* 2:147–149.

Hayes, P.A., and J.B. Cohen. 1987. *Further development of a mathematical model for the specification of immersion clothing insulation.* RAF, IAM Report 653.

Hayward, J.S., Eckerson, J.D., and M.L. Collis. 1975. Thermal balance and survival time prediction of man in cold water. *Canadian Journal of Physiology & Pharmacology* 53:21–32.

Hayward, J.S., Eckerson J.D., and M.L. Collis. 1975. Effect of behavioural variables on cooling rate of man in cold water. *Journal of Applied Physiology* 38:1073–1077.

Hayward, J.S., Eckerson, J.D., and D. Kemna. 1984. Thermal and cardiovascular changes during three methods of resuscitation from mild hypothermia. *Resuscitation* 11:21–33.

Hayward, J.S., Lisson, P.A., Collis, M.L., and J.D. Eckerson. 1978. *Survival suits for accidental immersion in cold water: Design-concepts and their thermal protection performance.* University of Victoria Report, Victoria, B.C.

Healiss, R. 1955. *Adventure glorious.* London: Frederick Muller Ltd.

Hensel, H. 1981. Thermoreception and temperature regulation. *Monographs of the Physiological Society* No. 38. Academic Press.

Hermann, R., and A. Stormer. 1985. Results of recent research of lifejackets. *Journal of the Royal Navy Medical Service* 3:161–166.

Hervey, G.R, and R.A. McCance. 1952. The effects of carbohydrate and seawater on the metabolism of men without food or sufficient water. *Proceedings of the Royal Society* B139:527–545.

Hervey, G.R., and R.A. McCance. 1954. Emergency rations. *Proceedings of Nutrition Society* 13:41–45.

Home Office. 1977. *Report of the working party on water safety.* HMSO. London.

Jessop, A. 1993. *Fatal accident inquiry.* Sherrifdom of Grampians, Highlands 7 Islands, Aberdeen, Scotland.

Keatinge, W.R. 1965. Death after shipwreck. *British Medical Journal* 2:1537–1541.

Keatinge, W.R. 1969. *Survival in cold water.* Oxford and Edinburgh: Blackwell Scientific.

Keatinge, W.R., and M. Evans M. 1960. Effect of food, alcohol, and hyoscine on body-temperature and reflex responses of men immersed in cold water. *Lancet* 2:176–178.

Keatinge, W.R., and J.A. Nadel.1965. Immediate respiratory response to sudden cooling of the skin. *Journal of Applied Physiology* 20:65–69.

Keys, A., Brozek, J., Henschel, A., Mickelsen, O., and H.L. Taylor. 1950. *The biology of human starvation.* Volumes I and II. Minneapolis: University of Minnesota Press.

Koffal. 1980. Personal correspondence.

Kvittingen, T.D., and A. Naess. 1963. Recovery from drowning in fresh water. *British Medical Journal* 1:1315–1317

Lancing, A. 1959. *Endurance: Shackleton's incredible voyage.* London: Hodder & Stoughton Ltd.

Laufman, H. 1951. Profound accidental hypothermia. *Journal of the American Medical Association* 147:1201–1212.

Leach, J. 1995. Psychological first-aid: A practical aide-memoire. *Aviation Space & Environmental Medicine* 66:668–674.

Lech, R.B. 1982. *All the drowned sailors.* Briarcliff Manor, New York: Stein & Day.

Lee, E.C.B., and K. Lee. 1971. *Safety and survival at sea.* London: Cassell, Giniger.

Lee, E.C.B., and K. Lee. 1989. *Safety and survival at sea.* New edition. London: Greenhill Books.

Lewis, M. 1999. Quoted in the *Scottish Sunday Post.* 21 February.

Lind, J. 1762. *An essay on the most effective means of preserving the health of seamen in the Royal Navy.* London: D. Wilson, Plato's Head Press.

Llano, G.A. 1955. *Airmen against the sea: An analysis of sea survival experiences.* Research Studies Institute, ADTIC Publication G-104, Maxwell Air Force Base, Alabama.

Lloyd, O.C. 1964. Cavers dying of cold. *Bristol Medical Chir. Journal* 79:1–5.

Lotens, W.A. 1993. *Heat transfer from humans wearing clothing.* Soesterberg, The Netherlands: TNO-Institute for Perception.

MacNalty, A.S. 1968. *Medical services in war: Rescue of survivors at sea.* Volume II. 9–12. HMSO. London.

Marriott, B.M. 1995. *Not enough to eat.* Washington, D.C.: National Academy Press.

Martineau, L., and I. Jacobs. 1989. Muscle glycogen availability and temperature regulation in humans. *Journal of Applied Physiology* 66:72–78.

Matthes, M. 1946. German air-sea rescue services: World War II. In *German aviation medicine World War II,* 1148. Surgeon General USAF, Department of the Air Force.

McArdle, W.D., Katch, F.I., and V.L. Katch. 1996. *Exercise physiology.* Baltimore: Williams & Wilkins.

McArdle, W.D., Magel, J.R., Gergley, T.J., Spina, R.J., and M.M. Toner. 1984. Thermal adjustment to cold-water exposure in resting men and women. *Journal of Applied Physiology.* 56:1565–1571.

McArdle, W.D., Magel, J.R., Spina, R.J., Gergley, T.J., and M.M. Toner. 1984. Thermal adjustments to cold-water exposure in exercising men and women. *Journal of Applied Physiology* 56:1572–1577.

McCance, R.A., Ungley, C.C., Crossfill, J.W.L., and E.M. Widdowson. 1956. *The hazards to men in ships lost at sea, 1940–44.* Medical Research Council, Special Report Series No. 291. HMSO. London.

McConnell, D.H., Brazier, J.R., Cooper, N., and C.D. Buckberg. 1977. Studies of the effects of hypothermia on regional myocardial blood flow and metabolism during cardiopulmonary by-pass. *Journal of Thoracic & Cardiovascular Surgery* 73:95–101.

Mekjavic, I.B., and J. Bligh. 1987. The pathophysiology of hypothermia. *International Reviews on Ergonomics* 1:201–218.

Mekjavic, I.B., Tipton, M.J., Gennser, M., and O. Eiken. 2001. Motion illness increases the risk of hypothermia. *Journal of Physiology*. 532.2:619–623.

Modell, J.H. 1971. *Pathophysiology and treatment of drowning*. Springfield, Illinois: Charles C Thomas.

Modell J.H., Graves, S.A., and A. Ketover. 1976. Clinical course of 91 consecutive near-drowning victims. *Chest* 70: 231-238.

Mohri, H., Dillard, D.H., Crawford, E.W., Martin, W.E., and K.A. Merendino. 1969. Method of surface induced deep hypothermia for open-heart surgery in infants. *Journal of Thoracic Cardiovascular Surgery* 58:262–270.

Molnar, G.W. 1946. Survival of hypothermia by men immersed in the ocean. *Journal of the American Medical Association* 131:1046–1050.

Money, K.E. 1970. Motion sickness. *Physiology Reviews* 50:1–39.

Moritz, A.R. 1944. Chemical methods of determination of death by drowning. *Physiology Reviews* 24:70–88.

Mundle, R. 1999. *Fatal Storm: the 54th Sydney to Hobart yacht race*. Australia: Harper Collins.

Naucler, S. 1757. Berichte von einem Mannes welcher dem Anschein nach efroren war denn aber wieder zum Leben verhollen war. *K. Schwed. Akad. Wiss.* 18:107.

Newburgh, L.H. 1968. *Physiology of heat regulation and the science of clothing*. New York and London: Hafner.

Nicholl, G.W.R. 1960. *Survival at sea: The development, operation and design of inflatable marine lifesaving equipment*. London: Adlard Coles Ltd. and New York: John De Graff.

Nunnely, S.A., and W.H. Wissler. 1980. *Prediction of immersion hypothermia in men wearing anti-exposure suits and/or using liferafts*. AGARD-CP-286, A1-1-A1-8.

Oakley, E.H.N. 1997. *Survival in a disabled submarine: Local cold injury and other medical problems*. Institute of Naval Medicine Report No. 97031, U.K.

Oakley, E.H.N., and R.J. Pethybridge. 1997. *The prediction of survival during cold immersion: Results from the UK National Immersion Incident Survey*. Institute of Naval Medicine Report No. 97011, U.K.

Orlowski, J.P. 1988. Drowning, near-drowning, and ice water drowning. *Journal of the American Medical Association* 260:390–391.

Ortzen, L. 1974. *Stories of famous shipwrecks*. London: Arthur Barker, Ltd.

Parsons, K.C. 1993. *Human thermal environments: The principles and the practice*. London: Taylor & Francis.

Passias, T.C., Meneilly, G.S., and I.B. Mekjavic. 1996. Effect of hypoglycaemia on thermoregulatory responses. *Journal of Applied Physiology* 80:1021–1032.

Philbrick, N. 2000. *In the heart of the sea*. New York: Harper Collins.

Pugh, L.G.C., and O.G. Edholm. 1955. The physiology of channel swimmers. *Lancet* 2:761–768.

Reason, J.T., and J.J. Brand. 1975. *Motion sickness*. London: Academic Press.

Reinke, J.J. 1875. *Beobachtungen uber die Koerpertemperaturbetrunkoner*. Deutsches Archives Fuer Kun. Medizin.

Robert Gordon Institute of Technology. 1988. *In-water performance assessment of lifejacket and immersion suit combinations.* Department of Energy Report. OTI 88 538.

Robertson, D. 1973. *Survive the savage sea.* London: Elek Books Ltd.

Robertson, D. 1975. *Sea survive: A manual.* London: Elek Books Ltd.

Robertson, D.M., and M.E. Simpson. 1995. *Review of probable survival times for immersion in the North Sea.* (UK) HSE Offshore Technology Report. OTO 95 038.

Robin, B. 1981. *Survival at sea.* London, Melbourne, Sydney, Auckland, and Johannesburg: Stanley Paul.

Rowell, L.B. 1986. *Human circulation. Regulation during physical stress.* New York: Oxford University Press.

Sargeant, A.J. 1987. Effect of muscle temperature on leg extension force and short-term power output in humans. *European Journal of Applied Physiology* 56:693–698.

Sharpey-Schafer, E.P., Hayter, C.J., and E.D. Barlow. 1958. Mechanism of acute hypotension from fear or nausea. *British Medical Journal* 2:878–880.

Simpson, A.W.B. 1994. *Cannibalism and the common law: A Victorian yachting tragedy.* London: Hambledon Press.

Siple, P.A., and C.F. Passel. 1945. Dry atmospheric cooling in subfreezing temperatures. *Proceedings of the American Philosophical Society* 89:177–199.

Smith, F.E. 1976. *Survival at sea.* Report prepared for the Survival at Sea Subcommittee of the Royal Naval Personnel Research Committee. Medical Research Council.

Steinman, A.M., Hayward, J.S., Nemiroff, M.J., and P.S. Kublis. 1987. Immersion hypothermia: comparative protection of anti-exposure garments in calm versus rough seas. *Aviation & Space Environmental Medicine* 58:550–558.

Tikuisis, P. 1994. In *Proceedings of the Sixth International Conference on Environmental Ergonomics,* ed. J. Frim, M.B. Ducharme, and P. Tikuisis, 160–1. Montebello.

Tikuisis, P. 1997. Prediction of survival time at sea based on observed body cooling rates. *Aviation Space & Environmental Medicine* 68:441–448.

Tikuisis, P., Gonzalez, R.R., and K.B. Pandolf. 1988. Thermoregulatory model for immersion of humans in cold water. *Journal of Applied Physiology.* 64:719–727.

Time magazine. 1994. The cruel sea. 10 October, 26–31.

Tipton, M.J. 1989. The initial responses to cold water immersion in man. *Clinical Science* 77:581–588.

Tipton, M.J. 1995. Immersion fatalities: Hazardous responses and dangerous discrepancies. *Journal of the Royal Naval Medical Service* 81:101–107.

Tipton, M.J., and P.J. Balmi. 1996. The effect of water leakage on the protection provided by immersion protective clothing worn by man. *European Journal of Applied Physiology* 72:394–400.

Tipton, M.J., Eglin, C., Gennser, M., and F. Golden. 1999. Immersion deaths and deterioration in swimming performance in cold water. *Lancet* 354:626–629.

Tipton, M.J., Franks, C.M., Meneilly, G.S., and I.B. Mekjavic. 1997. *Estimation of diver survival time in a lost bell.* HSE Offshore Technology Report. OTH 96 516. Norwich, United Kingdom: HSE Books.

Tipton, M.J., Kelleher, P.C., and F.St.C. Golden. 1994. Supraventricular arrhythmias following breath-hold submersions in cold water. *Undersea and Hyperbaric Medicine* 21:305–313.

Toner, M.M., Sawka, M.N., and K.B. Pandolf. 1984. Thermal responses during arm and leg and combined arm-leg exercise in water. *Journal of Applied Physiology* 56:1355–1360.

Ungley, C.C., Channell, G.D., and R.L. Richards. 1945. The immersion foot syndrome. *British Journal of Surgery* 33:17–31.

Veghte, J.H. 1972. Cold sea survival. *Aerospace Medicine* 43:506–511.

Veicsteinas, A., Ferretti, G.T., and D.W. Rennie. 1982. Superficial shell insulation in resting and exercising man in cold water. *Journal of Applied Physiology* 52:1557–1564.

Wayburn, E. 1947. Immersion hypothermia. *Archives Internal Medicine* 79:77–91.

Windle, C.M. 1998. *Recommended improvements to disabled submarine survival rations.* Institute of Naval Medicine Report, No. 98020, U.K.

Wissler, W.H. 1981. A mathematical model of the human thermal system with reference to diving. In *Thermal constraints in diving,* ed. L.A. Kuehn, 187–212. 24th Undersea Medical Society Workshop, Bethesda.

Wolf, A.V. 1958. *Thirst—Physiology of the urge to drink and the problems of water lack.* Springfield, Illinois: Charles C Thomas.

Young, A.J., Sawka, M.N., Neufer, P.D., Muza, S.R., Askew, E.W., and K.B. Pandolf. 1989. Thermoregulation during cold water immersion is unimpaired by low muscle glycogen levels. *Journal of Applied Physiology* 66:1809–1816.

Index

Note: The italicized *f*, *t*, and *n* following page numbers refer to figures, tables, and footnotes, respectively.